MINING SEQUENTIAL PATTERNS FROM LARGE DATA SETS

T0137770

The Kluwer International Series on
ADVANCES IN DATABASE SYSTEMS

Series Editor
Ahmed K. Elmagarmid

Purdue University
West Lafayette, IN 47907

Other books in the Series:

For a complete listing of books in this series, go to http://www.springeronline.com

MINING SEQUENTIAL PATTERNS FROM LARGE DATA SETS

by

Wei Wang
University of North Carolina at Chapel Hill
NC, USA

Jiong Yang
Case Western Reserve University
Cleveland, OH, USA

 Springer

Wei Wang
University of North Carolina at Chapel Hill
Chapel Hill, NC, USA

Jiong Yang
Case Western Reserve University
Cleveland, OH, USA

Library of Congress Cataloging-in-Publication Data

A C.I.P. Catalogue record for this book is available
from the Library of Congress.

MINING SEQUENTIAL PATTERNS FROM LARGE DATA SETS
by Wei Wang
University of North Carolina at Chapel Hill, NC, USA
Jiong Yang
Case Western Reserve University, Cleveland, OH, USA

The Kluwer International Series in Advances in Database Systems Volume 28

ISBN 978-1-4419-3707-0   e-ISBN 978-0-387-24247-7

Printed on acid-free paper.

Printed in the United States of America.

9 8 7 6 5 4 3 2 1

springeronline.com

Contents

List of Figures

List of Tables

Preface

In many applications, e.g., bioinformatics, web access traces, system utilization logs, etc., the data is naturally in the form of sequences. It has been of great interests to analyze the sequential data to find their inherent characteristics. The sequential pattern is one of the most widely studied models to capture such characteristics. Examples of sequential patterns include but are not limited to protein sequence motifs and web page navigation traces.

In this book, we focus on sequential pattern mining. To meet different needs of various applications, several models of sequential patterns have been proposed. We do not only study the mathematical definitions and application domains of these models, but also the algorithms on how to effectively and efficiently find these patterns.

The objective of this book is to provide computer scientists and domain experts such as life scientists with a set of tools in analyzing and understanding the nature of various sequences by : (1) identifying the specific model(s) of sequential patterns that are most suitable, and (2) providing an efficient algorithm for mining these patterns.

Chapter 1

INTRODUCTION

Data Mining is the process of extracting implicit knowledge and discovery of interesting characteristics and patterns that are not explicitly represented in the databases. The techniques can play an important role in understanding data and in capturing intrinsic relationships among data instances. Data mining has been an active research area in the past decade and has been proved to be very useful.

In many applications, data typically is represented in the form of sequence(s). This may be either a consequence of employing a natural temporal ordering among individual data (e.g., financial market data) or a result of complying to some inherent physical structure (e.g., genes in chromosomes). The input sequence is usually very long, which demands high scalability of the pattern discovery process. As an important research direction of the data mining field, mining patterns in long sequential data has been widely studied during recent years, which includes but is not limited to the following application domains.

- *Bio-Medical Study.* Each human gene consists a sequence of (usually over a half million) nucleotide acids. A certain combination of nucleotide acids may uniquely define a specific aspect of the biological function or appearance of a cell. Identifying such kind of meaningful combinations plays a crucial role in understanding the fundamental of life towards a deeper level, which has been the goal of many biologists for many years. By viewing the chromosomes as sequences of nucleotide acids, the above task can be transformed into mining sequential patterns that satisfying some user-specified criteria.

- *Performance Analysis.* Many system-monitoring applications involve collecting and analyzing attributes whose values evolve over time. Patterns of system state transition have been proved to be very useful in predicting system behavior from a recent state history and in preventing disastrous circumstances from occurring.

- *Client Profile.* User profiles can be built based on the discovered patterns on trace logs. Such knowledge can either be utilized to develop an optimal proxy caching scheme or be used to provide better market targeting tools.

- *Consumer Behavior.* Most retailers have customer rewarding programs that give consumers incentives to use their VIP cards when they shop. This allows to monitor the sequence of shopping events for each consumer. Patterns in this collection of transactions will be very helpful in revealing consumers' purchase affinity.

As a matter of fact, noises exist in most applications, which adds considerable challenges to the pattern mining process simply because many important patterns may be concealed if the model employed fails to accommodate noises properly. Depending on the type of application and the user's interests, tolerable noises may present in different formats and hence require different models accordingly.

1 *Injection of noise.* A typical example of noise injection is that, a client may accidentally access some irrelevant web page by mistake when he/she surfs on the Internet. Such access entries may be regarded as random noises inserted in the long traces during the process of mining meaningful patterns from the collected trace logs.

2 *Over-population of uninteresting patterns.* Different symbols (or events) may occur at vastly different frequencies by nature. For example, the sales of lamps is typically much higher than that of big furniture. However, even though patterns involving less expansive goods (such as lamps) occur more frequently, they may be considered not very interesting if the occurrences of such patterns are within people's expectation. In contrast, unexpected patterns involving furniture, if any, may be of great interests (probably because of a considerably high profit margin) even though such patterns may have relatively small number of occurrences in the data. Unfortunately, the significance of an infrequent but interesting pattern can be easily diluted by the "over-populated" unwanted patterns. These unwanted patterns can be viewed as "noise" in a broad sense because their presence *obstructs* (to some extent) the discovery of interesting infrequent patterns.

3 *Noise in the form of substitution.* It is possible that, in the presence of noise, a symbol is misrepresented as another symbol in the observation. Well-known examples include mutations between amino acids in proteins. This symbol substitution may prevent an occurrence of a pattern from being

recognized in the mining process, and this impact becomes more severe for long patterns.

To address above issues, powerful model(s) that suits each specific purpose is demanded to provide clear separation between useful patterns and noises, and efficient mining algorithms are also needed to make these new models applicable to long data sequences. As mentioned before, the sequence can be very long and can easily range to more than hundreds of millions of symbols. This may result in very long patterns that could contain thousands of symbols. Therefore, any pattern discovery algorithm has to scale well with respect to both the length of the input sequence and the length of potential patterns. In this book, we focus on each type of noise in a separate chapter and present some recent advances to meet the challenges. In Chapter 2, we provide an overview of related work of the subject presented in this book. Chapters 3, 4, and 5 are devoted to recent advances in mining sequential patterns. Chapter 3 presents two recently proposed models, namely asynchronous patterns [5] and meta patterns [2], to address the issues of accommodating insertion of random noises and characterizing change of behavior. Chapter 4 discusses two methods, InfoMiner [3] and STAMP [4], for mining interesting patterns involving infrequent event(s) in long data sequences. Chapter 5 describes recent research [6, 1] on handling noise mainly in the form of substitutions. Finally, a conclusion remark is drawn in Chapter 6.

References

[1] Kum, H., Pei, J., Wang, W., and Duncan, D. (2003). ApproxMAP: approximate mining of consensus sequential patterns. *Proc. of SIAM Int'l. Conference on Data Mining (SDM)*.

[2] Wang, W., Yang, J., and Yu, P. (2001) Meta-patterns: revealing hidden periodic patterns. *IEEE Int'l. Conference on Data Mining (ICDM)*. pp. 550-557.

[3] Yang, J. Wang, W., and Yu, P. (2001). Infominer: mining surprising periodic patterns. *Proc. of the Seventh ACM Int'l Conference on Knowledge Discover and Data Mining (KDD)*. pp. 395-400.

[4] Yang, J. Wang, W., and Yu, P. (2003). STAMP: on discovery of statistically important pattern repeats in long sequential data. *Proc. of the Third SIAM International Conference on Data Mining (SDM)*.

[5] Yang, J., Wang, W., and Yu, P. (2003). Mining asynchronous periodic patterns in time series data. *IEEE Transactions on Knowledge and Data Engineering*. 15(3):613-628.

[6] Yang, J., Wang, W., Yu, P., and Han, J. (2002). Mining long sequential patterns in a noisy environment. *Proc. ACM SIGMOD Int'l. Conference on Management of Data (SIGMOD)*. pp. 406-417.

Chapter 2

RELATED WORK

Sequential pattern discovery has been an active research topic in past several years. The application of sequential pattern mining spans over a wide range, from analyzing user access patterns of a web site to the protein motif discovery, from studying the workload of a large computer system to the child-abuse cases, etc. The diversity of the applications suggest that it may not be possible to apply a single sequential pattern model to all these problems. Each application may require a unique model and solution. A number of research projects were established in recent years to develop meaningful sequential pattern models and efficient algorithms for mining these patterns. Most of these models belong to one of the following four categories, frequent patterns, periodic patterns, statistically significant patterns, and approximate patterns. We will describe some of the state of art achievement within these fields in this chapter.

1. Frequent Patterns

In many applications, the frequency can be viewed as a very useful metric to indicate the importance of a pattern. If a pattern occurs a large number of times in a data set, then this pattern may be important in characterizing or analyzing the data set. For instance, in a long genome (DNA) sequence, a pattern that occurs a large number of times may correspond to a tandem repeat, which can be very interesting to a molecular biologist. In the application of intrusion detection, there may be a set of sequences each of which represents the system calls of an intrusion process. A pattern that occurs in a large percentage of these sequences may be a signature for the intrusion.

The frequent pattern is one of earliest problems studied in the realm of the sequential pattern mining. It was first studied in [2, 19]. The input data is a set of sequences, where each sequence is a list of sets. For instance, a sequence can be in the form of $\{a\}\{b,c\}\{d\}$ where the first occurred set of items is a, the secondly occurred set of items are b and c and etc. b and c can be considered as occurring simultaneously. We say that a sequence S supports a pattern P if S contains P (or P appears in S). In the above example, $\{a\}\{b,c\}\{d\}$ supports pattern $\{a\}\{b\}$, $\{b,c\}\{d\}$, and $\{a\}\{d\}$. The support of a pattern P in a data set is the percentage of the sequences that support P. These patterns whose support is above a user-defined threshold are called frequent sequential patterns. The goal of frequent sequential pattern mining is to find all frequent sequential patterns from a given data set with a user specific support threshold.

The authors in [1] used a technique similar to that of level-wise mining algorithm of the frequent itemset mining for discovering frequent sequential patterns. The main idea behind the Apriori algorithm is that if a longer pattern P is qualified as a frequent pattern, any of P's sub-patterns should also be qualified as a frequent pattern, which is known as the Apriori property. Formally, the Apriori property can be described as the following.

PROPERTY 1.1 **Apriori Property.** *Let Π be a set of sequences, P be a sequential pattern, and $supp(P)$ be the support of P in Π. We have, $supp(P) \leq \min_{P' \subset P}(supp(P'))$ where P' is a sub-pattern (subsequence) of P.*

The Apriori property can be directly derived from the definition of support. In order for a pattern P to appear in a sequence S, every sub-pattern of P has to appear in S. This means that S supports every sub-pattern of P, and in turn, the support of P is less than or equal to that of every sub-pattern of P.

Based on the Apriori property, the authors in [1] devise a level-wise search algorithm. The algorithm proceeds one level at a time. In level i, it finds i-patterns (i.e., patterns with i positions or sets). In the first level, the algorithm finds the frequent patterns with only one set. From these 1-patterns, it constructs the candidate set of 2-patterns in the following manner. If $\{a\}$ and $\{b,c\}$ are the frequent 1-patterns, then it is possible that $\{a\}\{b,c\}$ and $\{b,c\}\{a\}$ could be frequent 2-patterns. Next, the actual support of the 2-patterns are computed based on the entire data set. The algorithm terminates at level j when there is no any frequent pattern with j sets found.

However, this method may not be efficient if the pattern is very long, ie., consists of a large number of positions. To further improve the performance, projection-based algorithms such as PrefixSpan [16] was introduced to reduce the candidate patterns and hence reduce the number of scans of the data. The main technique of projection-based pattern mining is to search the patterns in a depth-first fashion. After discovering pattern P as a frequent pattern, the algorithm searches the patterns with P as the prefix. For instance, if pattern a is deemed frequent, next, pattern $\{a\}\{b\}$, $\{a\}\{c\},\dots$ will be searched. Due to the order of pattern search, the data set can be partitioned in the following manner. To compute the support of patterns having P as the prefix, we only need to search in the sequences that contain P as a subsequence. Therefore, during the depth-first search, we can recursively partition (project) the data sets according to the prefixes. As a result, the search time can be significantly reduced.

2. Regular Patterns

The above frequent sequential pattern mining algorithms discover the sequential patterns that occur many times in a set of sequences. However, it does not care the position where a pattern occurs. In some application, a user may not be interested in the frequent patterns but rather be interested in the patterns that occur in some regularity. For instance, if we can find sequential patterns that occur in some periodicity, then we can predict the occurrences of some event in the future and better understand the inherent characteristics of the underlying data set.

There are two kinds of regularity. One is the cyclic patterns and the other is periodic patterns. The cyclic pattern was proposed in [13], which is also called cyclic association rules. The input data to [13] is a set of transactions, each of which consists of a set of items. In addition, each transaction is tagged with an execution time. The goal is to find association rules that repeat themselves throughout the input data. An associate rule may exhibit a cyclic behavior which can be represented as (l, o). An association rule has a cycle (l, o) if the association rule holds in every lth time unit starting at time unit o. For instance, if the unit of time is an hour and "coffee $\rightarrow$ dough-nuts" holds during the interval 7AM to 8AM every day (i.e., every 24 hours), the rule "coffee $\rightarrow$ dough-nuts" has a cycle $(24, 7)$. The authors proposed two methods for discovering the cyclic patterns. The main ideas behind these approaches is to use some pruning techniques to reduce the computation. For instance, if itemset A and B have different non-overlapped cycles, then it is impossible for the rule $A \rightarrow B$ to exhibit cyclic behavior.

For the periodic patterns, the input is a long sequence of sets of items. The goal is to find the subsequences that exhibit the periodicity in the input sequence. Assume that $\{a\}\{b\}\{c\}\{a\}\{b\}\{c\}\{a\}\{b\}\{c\}$ is the input sequence. The pattern $\{a\}\{b\}\{c\}$ is called periodic pattern because it repeats itself with period 3. This is also called full periodic pattern because every position in the pattern exhibits the periodicity. However, in many applications, not every position may exhibit the periodic behavior. For example, in a set of custom purchase transactions, a periodic trend may be only observed at certain time of a day. For instance, let $\{a\}\{a\}\{c\}\{a\}\{b\}\{c\}\{a\}\{c\}\{c\}$ be the input sequence, then there is no full periodic pattern with length 3. However, when the constraint is relaxed, e.g., some position can be don't care, then we can find the pattern $\{a\} * \{c\}$ where $*$ is a wide card and can represent any set of items. This is called *partial periodic pattern*.

Han et. al. [9] presented algorithms for efficiently mining these partial periodic patterns by exploring some interesting properties related to partial periodicity such as the Apriori property and the max-sub-pattern hit set property. In essence, to discover patterns with periodicity of l, the algorithm divides input sequence into a set of contiguous segments with length l, each of which can be viewed as an independent sequence, then a frequent sequential pattern mining algorithm is used. Although the partial periodic pattern model can capture the patterns where some position is non-specified, it could not represent the patterns whose occurrences are asynchronous. For instance, if there is some noise in the input sequence and some sets are missed or extra sets are added (which is a common phenomenon in many real applications), e.g., the input sequence may become $\{a\}\{a\}\{c\}\{a\}\{c\}\{a\}\{c\}\{c\}$ due to noises, then there does not exist any partial periodic pattern with periodicity of 3.

3.   Statistically Significant Patterns

In many applications, the symbols in a sequence have very skewed distribution. For instance, in a trace of messages of some router, some message type occurs very rarely, e.g., "the link connected to the router is saturated" while other type of messages may occur commonly, e.g., "I am alive" message. As a result, the patterns with common messages should be expected to have a higher support than that of these with the rare messages. If we use the uniform support (number of occurrences) as a measure of importance, then we would miss these patterns with rare events. Researchers have been investigating this problem in various data mining applications.

Brin et al. [4] first introduced the concept of correlation and it was shown that in many applications the correlation measurement can reveal some very important patterns. The Chi-squared test was used to test the correlation among items. For example, by analyzing the words in the *clari.world.africa* news article, Brin et. al. found that there exist some very strong correlation among some words combination, e.g., "nelson" and "mandela", but their support is relatively low. In this approach, the itemsets form a lattice based on the superset-subset relationship. Instead of explicitly enumerating all correlated itemsets, the border comprising the set of minimal correlated itemsets[1] is identified, and no further distinction is made on the degree of correlation of itemsets above the border (i.e., supersets of some itemset on the border). This model sometimes becomes sub-optimal. As shown in Figure 2.1, itemsets A and B are highly correlated but C is independent of them[2]. In addition, $\{A, B, C, D\}$ is also highly correlated. We can view that the degree of correlation of $\{A, B, C\}$ is not as strong as that of $\{A, B\}$ and $\{A, B, C, D\}$. This observation can also be confirmed by the Chi-squared test[3]. In many applications, users are only interested in the itemsets such as $\{A, B\}$ and $\{A, B, C, D\}$, but not $\{A, B, C\}$. However, [4] cannot distinguish between $\{A, B, C\}$ and $\{A, B, C, D\}$ once $\{A, B\}$ is

Transaction ID	Items
1	ABCD
2	ABFG
3	CEGF
4	ABCD
5	CEGH
6	CEFH

Figure 2.1.   An Example of Transaction Set

identified as a correlated itemset. Furthermore, if a user is interested in finding k itemsets with the highest correlation, then all itemsets in the lattice have to be examined before k highest ones can be determined. Another potential drawback of this model is the expensive computation required by this model. The running time of all patterns with i-correlated items is $O(n \times |CAND| \times \min\{n, 2^i\})$ where n and $|CAND|$ are the number of transactions and the number of candidates at the ith level, respectively. To overcome these drawbacks, Oates et al. [11, 12] proposed models for statistical dependencies using G statistic and devised randomized algorithms to produce approximate results.

More recently, Cohen et al. [6] and Fujiwara et al. [8] address the problem of identifying pairs of attributes with high confidence or similarity (in terms of probabilistic correlations in the database) in the absence of support requirement. Hashing based algorithms [6] are proposed to tackle the problem, which consist of three general phases: computing hash signature, generating candidates, and pruning candidates. To avoid both false negatives and false positives that may be yielded with the hashing based scheme, a family of so called *dynamic miss counting* algorithms are proposed in [8]. Instead of counting the number of hits (as most other algorithms do), the number of transactions where the given pair of attributes disagree is counted and this counter is deleted as soon as the number of misses exceeds the maximum number of allowed misses for that pair. This strategy is proved to be able to reduce the memory size significantly.

Another important advance is accomplished in mining so-called *unexpected patterns*. Berger et al. [3] proposed a probabilistic measure of interestingness based on unexpectedness in the context of temporal logic, whereby a pattern is deemed interesting if the ratio of the actual number of occurrences of the pattern exceeds the expected one by some user defined threshold. Solving the problem in its general frame is in nature NP-hard and hence some heuristics are proposed to produce an approximate answer. Padmanabhan et al. [14, 15, 18] define the unexpectedness of association rules relative to a system of prior beliefs. Specifically, the belief is of the form $X \to Y$ and a rule is said to be unexpected if it contradicts the belief. The set of beliefs (given by the user) are used to conduct the mining process efficiently so that an exhaustive search is avoided. The primary advantage of this model is that it can customize the mining process for the users who have fairly good prior knowledge and specific interests, and is particularly useful in refinements of user's beliefs.

A formal study of surprising patterns is furnished in [5], focusing on the analysis of variation of inter-item correlations along time. The surprise is defined in terms of the coding length in a carefully chosen encoding scheme and has solid theoretic foundation, but requires much more expensive computation comparing to other models.

4. Approximate Patterns

Approximate frequent itemset mining is studied in [17, 20]. Although the two methods are quite different in techniques, they both explored approximate matching among itemsets. Instead of perfect matchs, the model in [17, 20] allow imperfect matchs. A transaction supports an itemset I if a large portion of the items in I, e.g., more than 95% occur in the transaction. Algorithms similar to mining perfect itemsets are devised to mine the approximate itemsets.

In [7], Chudova and Smyth used a Bayes error rate framework under a Markov assumption to analyze different factors that influence string pattern mining in computational biology. Extending the theoretical framework to mining sequences of sets could shed more light to the future research direction.

Notes

1   A minimal correlated itemset is a correlated itemset whose subsets are all independent.
2   $Prob(AB) \times Prob(C) = \frac{1}{2} \times \frac{2}{3} = Prob(ABC)$.
3   In general, the chi-squared test requires a large sample. For the demonstration purpose only, we assume that the chi-squared test is valid in this example.

References

[1] Agrawal, R., and Srikant, R. (1994). Fast algorithms for mining association rules in large databases. *Proc. of the Int'l Conference on Very Larg Databases (VLDB).* pp. 487-499.

[2] Agrawal, R., and Srikant, R. (1995). Mininig sequential patterns. *Proc. of the Int'l Conference on Data Engineering (ICDE).* pp. 3-14.

[3] Berger, G., and Tuzhilin, A. (1998). Discovering unexpected patterns in temporal data using temporal logic. *Temporal Databases - Research and Practice, Lecture Notes on Computer Sciences.* vol. (1399) pp. 281-309.

[4] Brin. S., Motwani, R., and Silverstein, C. (1997). Beyond market baskets: generalizing association rules to correlations. *Proc. ACM SIGMOD Int'l. Conference on Management of Data (SIGMOD)*. pp. 265-276.

[5] Chakrabarti, S., Sarawagi, S., and Dom, B. (1998). Mining surprising patterns using temporal description length. *Proc. Int. Conf. on Very Large Data Bases (VLDB)*. pp. 606-617.

[6] Cohen, E., Datar, M., Fuijiwara, S., Cionis, A., Indyk, P., Motwani, R., Ullman, J., and Yang, C. (2000). Finding interesting associations without support pruning. *Proc. 16th IEEE Int'l. Conference on Data Engineering (ICDE)*. pp. 489-499.

[7] Chudova, D., and Smyth, P. (2002). Pattern discovery in sequences under a markov assumption. *Proc. of the Eighth ACM Int'l Conference on Knowledge Discover and Data Mining (KDD)*. pp. 153-162.

[8] Fujiwara, S., Ullman, J., and Motwani, R. (2000) Dynamic miss-counting algorithms: finding implication and similarity rules with confidence pruning. *Proc. 16th IEEE Int'l. Conference on Data Engineering (ICDE)*. pp. 501-511.

[9] Han, J., Dong, G., and Yin, Y. (1999) Efficient mining partial periodic patterns in time series database. *Proc. of IEEE Int'l. Conf. on Data Engineering*. pp. 106-115.

[10] Han, J., Pei, J., Mortazavi-Asl, B., Chen, Q., Dayal, U., and Hsu, M. (2000). FreeSpan: frequent pattern-projected sequential pattern mining. *Proc. of the Sixth ACM Int'l Conference on Knowledge Discover and Data Mining (KDD)*. pp. 355-359.

[11] Oates, T. (1999) Identifying distinctive subsequences in multivariate time series by clustering. *Proc. of the Fifth ACM Int'l Conference on Knowledge Discover and Data Mining (KDD)*. pp. 322-326.

[12] T. Oates, M. D. Schmill, and P. R. Cohen. (1999) Efficient mining of statistical dependencies. *Proc. 16th Int'l. Joint Conf. on Artificial Intelligence (JCAI)*. pp. 794-799.

[13] Ozden, B., Ramaswamy, S., and Silberschatz, A. (1998). Cyclic association rules. *Proc. 14th Int'l. Conference on Data Engineering (ICDE)*. pp. 412-421.

[14] Padmanabhan B., and Tuzhilin, A. (1998) A belief-driven method for discovering unexpected patterns. *Proc. of ACM Int'l Conference on Knowledge Discover and Data Mining (KDD)*. pp. 94-100.

[15] Padmanabhan, B., and Tuzhilin, A. (2000). Small is beautiful: discovering the minimal set of unexpected patterns. *Proc. of ACM Int'l Conference on Knowledge Discvoer and Data Mining (KDD)*. pp. 54-63.

[16] Pei, J., Han, J., Mortazavi-Asl, B., Pinto, H., Chen, Q., and Dayal, U., and Hsu, M. (2001). PrefixSpan: mining sequential patterns by prefix-projected growth. *IEEE Int'l Conference on Data Engineering (ICDE)*. pp. 215-224.

[17] Pei, J., Tung, A., and Han, J. (2001). Fault-tolerant frequent pattern mining: problems and challenges. *Proc. ACM SIGMOD Int'l Workshop on Data Mining and Knowledge Discovery (DMKD)*. pp. 7-12.

[18] Silberschatz, A., and Tuzhilin, A. (1996). What makes patterns interesting in knowledge discover systems. *IEEE Transactions on Knowledge and Data Engineering (TKDE)*. vol. 8 no. 6, pp. 970-974.

[19] Srikant, R., and Agrawal, R. (1996). Mining sequential patterns: generalizations and performance improvements. *Proc. of Int'l Conference on Extended Database Technologies (EDBT)*. pp. 3-17.

[20] Yang, C., Fayyad, U., and Bradley, P. (2001). Efficient discovery of error-tolerant frequent itemsets in high dimensions. *Proc. of ACM Int'l Conference on Knowledge Discover and Data Mining (KDD)*. pp. 194-203.

Chapter 3

PERIODIC PATTERNS

In this chapter, we discuss the problem of mining periodic patterns. Although some research has been conducted in this area, the periodic pattern model is quite rigid and it fails to find patterns whose occurrences are asynchronous. In addition, the previous periodic pattern models do not provide an overview on the evolution of patterns within the input sequence. We will describe two advanced models [8, 9] that can capture these types of periodic patterns and efficient algorithms for mining them in this chapter.

1. Asynchronous Patterns

Periodicity detection on time series data is a challenging problem of great importance in many real applications. Most previous research in this area assumed that the disturbance within a series of repetitions of a pattern, if any, would not result in the loss of synchronization of subsequent occurrences of the pattern with previous occurrences [3, 4]. For example, "Joe Smith reads newspaper every morning" is a periodic pattern. Even though Joe might not read newspaper in the morning occasionally, this disturbance will not affect the fact that Joe reads newspaper in the morning of the subsequent days. In other words, disturbance is allowed only in terms of "missing occurrences" but not as general as any "insertion of random noise events". However, this assumption is often too restrictive since we may fail to detect some interesting pattern if some of its occurrences is misaligned due to inserted noise events. Consider the application of *inventory replenishment*. The history of inventory refill orders can be regarded as a symbol sequence. Assume that the time between two replenishments of cold medicine is a month normally. The refill order is filed at the beginning of each month before a major outbreak of flu which in turn causes an additional refill at the 3rd week. Afterwards, even though the replenishment frequency is back to once each month, the refill time shifts to the 3rd week of a month (not the beginning of the month any longer). Therefore, it would be desirable if the pattern can still be recognized when the disturbance is within some reasonable threshold. In addition, the system behavior may change over time. Some pattern may not be present all the time (but rather within some time interval). Therefore, in this section, we aim at the problem of mining periodic patterns that are significant within a subsequence of symbols which may contain disturbance of length up to a certain threshold [9]. Two parameters, namely *min_rep* and *max_dis*, are employed to qualify valid patterns and the symbol subsequence containing it, where this subsequence in turn can be viewed as a list of **valid segments** of perfect repetitions interleaved by disturbance. Each valid segment is required to be of at least min_rep contiguous repetitions of the pattern and the length of each piece of disturbance is allowed only up to max_dis. The intuition behind this is that a pattern needs to repeat itself at least a certain number of times to demonstrate its significance and periodicity. On the other hand, the disturbance between two valid segments has to be within some reasonable bound. Otherwise, it would be more appropriate to treat such disturbance as a signal of "change of system behavior" instead of random noise injected into some persistent behavior. The parameter max_dis acts as the boundary to separate these two phenomena. Obviously, the appropriate values of these two parameters are application dependent and need to be specified by the user. For patterns satisfying these two requirements, our model will return the subsequence with the maximum overall repetitions. Note that, due to the presence of disturbance, some subsequent valid segment may not be well synchronized with the previous ones. (Some position shifting occurs.) This in turn would impose a great challenge in the mining process.

Similar to [4], a pattern can be partially filled to enable a more flexible model. For instance, (*cold_medi*, *, *, *) is a partial monthly pattern showing that the cold medicine is reordered on the first week of each month while the replenishment orders in the other three weeks do not have strong regularity. However, since we also allow the shifted occurrence of a valid segment, this flexible model poses a difficult problem to be solved. For a give pattern P, its associated valid segments may overlap. In order to find the valid subsequence with the most repetitions for P, we have to decide which valid segment and more specifically which portion of a valid segment should be selected. While it is relatively easy to find the set of valid segments for a given pattern, substantial difficulties lie on how to assemble these valid segments to form the longest valid subsequence. As shown in Figure 3.1, with $min\_rep = 3$, S_1, S_2, and S_3 are three valid segments of the pattern $P = (d_1, *, *)$. If we set $max\_dis = 3$, then X_1 is the longest subsequence before S_3 is considered, which in turn makes X_2 the longest one. If we only look at the symbol sequence up to position j without looking ahead in the sequence, it is very difficult to determine whether we should switch to S_2 to become X_1 or continue on S_1.

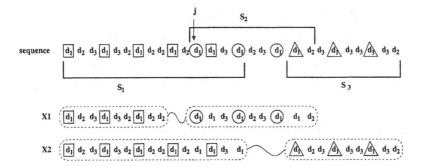

Figure 3.1. Example of Symbol Sequence

This indicates that we may need to track multiple ongoing subsequences simultaneously. Since the number of different assemblages (of valid segments) grows exponentially with increasing period length, the process to mine the longest subsequence becomes a daunting task (even for a very simple pattern such as $(d_1, *, *)$). To solve this problem, for a given pattern, an efficient algorithm is developed to identify subsequences that may be extended to become the longest one and organize them in such a way that the longest valid subsequence can be identified by a single scan of the input sequence and at any time only a small portion of all extendible subsequences needs to be examined.

Another innovation of our mining algorithm is that it can discover all periodic patterns regardless of the period length. Most previous research in this area focused on patterns for some pre-specified period length [3, 4, 5] or some pre-defined calendar [6]. Unfortunately, in practice, the period is not always available a priori (It is also part of what we want to mine out from the data). The stock of different merchandizes may be replenished at different frequencies (which may be unknown ahead of time[1] and may also varies from time to time). A period may span over thousands of symbols in a long time series or just a few symbols. We first introduce a distance-based pruning mechanism to discover all possible periods and the set of symbols that are likely to appear in some pattern of each possible period. In order to find the longest valid subsequence for all possible patterns, we employ a level-wise approach. The Apriori property also holds on patterns of the same period. That is, a valid segment of a pattern is also a valid segment of any pattern with fewer symbols specified in the pattern. For example, a valid segment for $(d_1, d_2, *)$ will also be one for $(d_1, *, *)$. Then, for each likely period, all valid patterns with their longest supporting subsequences can be mined via an iterative process.

Model of Asynchronous Periodic Patterns

In this section, we formally define the model of asynchronous periodic patterns. In short, we also call it asynchronous patterns. Let $\Im = \{d_1, d_2, \dots, \}$ be a set of literals and D be a sequence of literals in $\Im$. We first introduce some notations that would facilitate the discussion in the scope of asynchronous patterns.

Definition A **pattern** with **period** l is a sequence of l symbols $(p_1, p_2, \dots, p_l)$, where $p_1 \in \Im$ and the others are either a symbol in $\Im$ or *, i.e., $p_j \in \Im \cup *$ $(2 \le j \le l)$.

Since a pattern can start anywhere in a sequence, we only need to consider patterns that start with a non "*" symbol. Here, * is introduced to allow partial periodicity. In particular, we use * to denote the "don't care" position(s) in a pattern. A pattern P is called a i-**pattern** if exactly i positions in P are symbols from $\Im$. (The rest of the positions are filled by *.) For example, $(d_1, d_2, *)$ is a 2-pattern of period 3.

Definition For two patterns $P = (p_1, p_2, \dots, p_l)$ and $P' = (p_1', p_2', \dots, p_l')$ with the same period l, P' is a **specialization** of P (i.e., P is a **generalization** of P') iff, for each position $j(1 \le j \le l)$, either $p_j = p_j'$ or $p_j = *$ is true.

For example, pattern $(d_1, d_2, *)$ is considered as a specialization of $(d_1, *, *)$ and a generalization of (d_1, d_2, d_3).

Definition Given a pattern $P = (p_1, p_2, \dots, p_l)$ with period l and a sequence of l literals $D' = d_1, d_2, \dots, d_l$, we say that P **matches** D' (or D' **supports** P) iff, for each position $j(1 \le j \le l)$, either $p_j = *$ or $p_j = d_j$ is true. D' is also called a **match** of P.

In general, given a sequence of symbols and a pattern P, multiple matches of P may exist. In Figure 3.2(a), $D_1, D_2, \dots, D_7$ are seven matches of $(d_1, *, d_2)$. We say that two matches of the same period are **overlapped** iff they share some common subsequence, and are **disjoint** otherwise. For instance, D_1 and D_3 are disjoint while D_1 and D_2 are overlapped and their common subsequence is indicated by the shaded area in Figure 3.2(a).

Definition Given a pattern P with period l and a sequence of symbols D, a list of k ($k > 0$) disjoint matches of P in D is called a **segment** with respect to P iff they form a contiguous subsequence of D. k is referred to as the **number of repetitions** of this segment.

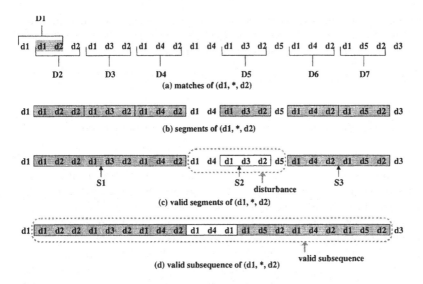

Figure 3.2. Example of matches and segments of $(d_1, *, d_2)$

Segments D_2, D_3, and D_4 form a contiguous subsequence as shown in Figure 3.2(b). Therefore, the subsequence $d_1, d_2, d_2, d_1, d_3, d_2, d_1, d_4, d_2$ is a segment with respect to the pattern $(d_1, *, d_2)$. Note that, by definition, each match of a pattern P itself is also a segment with respect to P.

Definition A segment S with respect to a pattern P is a **valid segment** iff the number of repetitions of S (with respect to P) is at least the required minimum repetitions (i.e., min_rep).

If the value of min_rep is set to 2, then both segments S_1 and S_3 qualify as valid segments as illustrated by shaded area in Figure 3.2(c). S_2 is not a valid segment since it only contains one match of $(d_1, *, d_2)$. In general, given a pattern P, any sequence of symbol can be viewed as a list of disjoint valid segments (with respect to P) interleaved by disturbance. For example, the subsequence enclosed in the dashed contour in Figure 3.2(c) is treated as disturbance between two valid segments S_1 and S_3.

Definition Given a sequence D and a pattern P, a **valid subsequence** in D is a set of non-overlap valid segments where the distance between any two successive valid segments does not exceed the parameter max_dis. The **over-**

all number of repetitions of a valid subsequence is equal to the sum of the repetitions of its valid segments. A valid subsequence with the most overall repetitions of P is called its **longest valid subsequence**.

Definition For a sequence of symbols D, if there exists some valid subsequence with respect to a pattern, then this pattern is called a **valid pattern**.

It follows from the definition that any valid segment itself is also a valid subsequence. If we set $max\_dis = 4$, even though S_1 and S_3 in Figure 3.2(c) are individually valid subsequences, there does not exist a valid subsequence containing both of them due to the violation of the maximum allowed disturbance between them. In contrast, the subsequence enclosed by dashed line in Figure 3.2(d) is a valid subsequence whose overall number of repetitions is 6.

For a given sequence of literals D, the parameters min_rep and max_dis, and the maximum period length L_{max}, we want to find the valid subsequence that has the most overall repetitions for each valid pattern whose period length does not exceed L_{max}.

Algorithm of Mining Asynchronous Periodic Patterns

In this section, we outline strategies to tackle the problem of mining subsequences with most overall repetitions for all possible patterns.

1 *Distance-based Pruning.* For each symbol d, we examine the distance between any two occurrences of d. Let $DC_{d,l}$ be the number of times when such distance is exactly l. For each period l, the set of symbols whose $DC_{d,l}$ counters are at least min_rep are retained for the next step.

2 *Single Pattern Verification.* For each potential period l and each symbol d that passed the previous step, a candidate 1-pattern $(d_1, d_2, \ldots, d_l)$ is formed by assigning $d_1 = d$ and $d_2 = \cdots = d_l = *$. We validate all candidate patterns $(d, *, \ldots, *)$ via a single scan of the sequence. Note that any single pattern of format $(*, \ldots, d, *, \ldots, *)$ is essentially equivalent to the pattern $(d, *, \ldots, *)$ (of the same period) with a shifted starting position in the sequence. A segment-based approach is developed so that a linear scan of the input sequence is sufficient to locate the longest valid subsequence for a given pattern.

3 *Complex Pattern Verification.* An iterative process is carried out where at the ith iteration, the candidate i-patterns are first generated from the set of valid $(i-1)$-patterns, and are then validated via a scan of the data sequence.

We now elaborate each step in following sections.

Distance-based Pruning of Candidate Patterns

Since there are a huge number of potential patterns, $O(|\Im|^{L_{max}})$, a pruning method is needed to reduce the number of candidates. The pruning method is motivated by our observation that if a symbol d participates in some valid pattern of period l, there should be at least min_rep times that the distance between two occurrences of d is exactly l (in order to form a valid segment). The proposed distance-based pruning method makes one pass over the data sequence to discover all possible periods and the set of symbols that are likely to appear in some pattern of each possible period. For each symbol d and period l, the number of times when the distance between two occurrences of d in the sequence is l is collected.

To perform the distance-based pruning, when scanning through the sequence, we need to maintain a moving window of the last L_{max} scanned symbols. For the next symbol, say d, we compare it with each of the symbol in the moving window. If a match occurs at the jth position, the count for period $(L_{max}-j+1)$ of symbol d (denoted as $DC_{d,L_{max}-j+1}$) is incremented by 1. For example, in Figure 3.2(a), the third d_1 in the fifth position will contribute to both $D_{d_1,3}$ and $D_{d_1,4}$. Due to the generality of our model, for each occurrence of a symbol d, we need to track its distance to all of its previous occurrences within the moving window. Our model does not only allow partially specified patterns such as $(d, *, *)$ where d can also occur in the "*" position, but also recognizes patterns with repetition of the same symbol such as $(d, *, d)$. Hence it is not sufficient to just track the distance of a symbol to its last occurrence. Given a symbol d and a period l, if $DC_{d,l}$ is larger than or equal to the min_rep threshold, then it is possible that d might participate in some valid pattern of period l. We can use this property to reduce the candidate patterns significantly.

Longest Subsequence Identification for a Single Pattern

If a symbol d and period l pair has passed the distance-based pruning, then the longest subsequence identification (LSI) algorithm is used to discover the subsequence with the most repetitions of $(d, *, \ldots, *)$ with period l. Each occurrence of d in the sequence corresponds to a match of the pattern. If d occurs at position i, then the subsequence from position i to position $(i+l-1)$ is a match of the pattern. Consider the pattern $(d_1, *, *)$ and the sequence in Figure 3.3(a). d_1 occurs 11 times, each of which corresponds to a match denoted by D_j $(1 \leq j \leq 11)$. Before presenting the algorithm, we first introduce the concept of *extendibility* and *subsequence dominance*.

Definition For any segment S with respect to some pattern P, let $D_1, D_2, \ldots, D_k$ be the list of matches of P that form S. Then the segment S' formed by

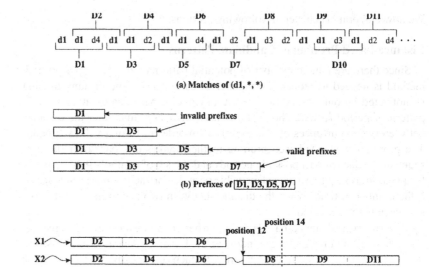

Figure 3.3.   Example of Extendibility and Subsequence Dominance

$D_1, \ldots, D_{k'}$ is called a **prefix** of S, where $1 \le k' \le k$. S is also referred to as an **extension** of S'. S' is called a **valid prefix** of S if S' is a valid segment.

Definition A segment S is **extendible** iff it is a prefix of some other segment S' $(S' \ne S)$.

Any segment is also a prefix of itself. For any segment S whose number of repetitions is x, there exist x different prefixes of S. Figure 3.3(b) shows all four possible prefixes of the segment $[D_1, D_3, D_5, D_7]$ in Figure 3.3(a), among which the first two are not valid prefixes while the other two are valid if $min\_rep = 3$. We also say that the first three prefixes in Figure 3.3(b) are extendible.

Definition For any two valid subsequences X and Y with the same starting position, X is a **prefix** of Y (and Y is an **extension** of X) iff each valid segment in X is also a valid segment in Y, except that the last valid segment S in X may be a prefix of a valid segment S' in Y. Let j be the starting position of the first match of the pattern in Y but not in X. Then we say that X is **extended**

on position j **to generate** Y.

Definition Given a pattern P, a valid subsequence X is **extendible** if there exists another valid subsequence Y ($Y \neq X$) such that Y is an extension of X.

In Figure 3.3(c), X_1, X_2, X_3 are three valid subsequences of $(d_1, *, *)$ if $min\_rep = 3$ and $max\_dis = 5$. X_1 is a prefix of X_2 and is extended at position 12 to generate X_2. As a result, X_1 is also said to be extendible.

Definition Given a position i, for any two valid subsequences X and Y that end between positions $(i - max\_dis - 1)$ and i, we say that X **dominates** Y at position i iff the overall number of repetitions of X is greater than or equal to that of Y.

It is clear that, at position 14 in Figure 3.3(c), X_3 dominates X_1. This subsequence dominance defines a total ordering among the set of valid subsequences that are considered to be *extendible* at any given position. If X dominates Y at some position i (Figure 3.4(a)), then the subsequence X' generated by appending Z to X starting at position i can not be shorter than the extension Y' of Y generated by appending Z to Y (Figure 3.4(b)). Therefore, we do not need to extend Y at position i. Note that we still have to keep Y since we may need to extend Y (to include V in Figure 3.4(c)) at some later position $j (j > i)$. This scenario may happen when X ends at an earlier position than Y does and X is known to be not extendible at position j. This provides the motivation and justification of the pruning technique employed in our algorithm.

Algorithm Description

When scanning through a sequence to determine the longest subsequence containing a pattern $(d, *, \ldots, *)$, the discovery process may experience the following phases repeatedly:

- Phase A: *Segment validation.* At least one instance of $(d, *, \ldots, *)$ is found (in the latest segment of a subsequence), but the number of repetitions of the pattern is still less than min_rep.

- Phase B: *Valid segment growth.* The segment is now considered to be valid and the repetition count may continue to grow.

- Phase C: *Extension* (or *disturbance*). A valid segment may have ended. It is now going through some disturbance or noise region to see whether it can get extended to another segment of the pattern within max_dis, referred

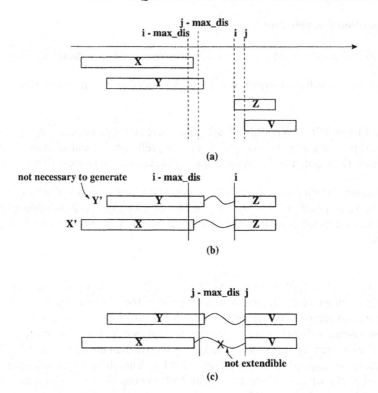

Figure 3.4.   Subsequence Dominance

to as the *extension window*. If so, it returns to Phase A. Otherwise the subsequence terminates.

There are several challenges here. First of all, the transition point from Phase B to Phase C is not obvious. Although missing the next consecutive match clearly signals the transition to Phase C, the presence of the next match does not necessarily mean the continuation of Phase B. This is illustrated by X_1 in Figure 3.1 at the $(j-2)$th position. Secondly, the transition point from Phase C to Phase A is not straightforward, either. In fact, any (not just the first) d occurring within the extension window can potentially be a candidate leading to a new extension of the subsequence. For X_2 in Figure 3.1, it is the second d_1 in the extension window that leads to a valid segment. We thus need to develop an efficient tracking mechanism for managing the phase transitions.

Furthermore, there is also the pruning issue. There can be many overlapping subsequences of a pattern. An efficient pruning criterion needs to be developed to prune the subsequences that cannot become the longest valid subsequence.

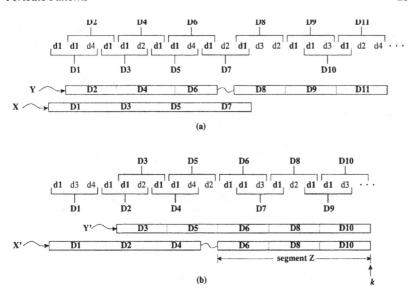

Figure 3.5. Valid subsequences of $(d_1, *, *)$ with $min\_rep = 3$ and $max\_dis = 5$

This will reduce the number of concurrent subsequences that need to be tracked. The problem here is that the longest subsequence at a particular instant may be overtaken by a shorter overlapping subsequence. This is clearly demonstrated in Figure 3.5(a) by X and Y. Assume that $min\_rep = 3$ and $max\_dis = 5$. After recognizing segment D_7 (before D_8 is encountered), X overtakes Y even though Y grows to be the longest valid subsequence later. However, we also observe that for any two valid subsequences X' and Y', if X' begins to dominate Y' at some position k in the sequence, any further extension of Y' will always be dominated by some extension of X' (Figure 3.5(b)). We can thus prune Y' after position k. A good point to check for dominance relationship is at a point when a segment Z first becomes valid. This is the point where X' that encompasses Z becomes valid and the two overlapping subsequences converge to a common tail segment.

Inspired by above observations, the algorithm can be outlined as follows. As scanning through the input data sequence, each time a match of pattern $(d, *, \ldots, *)$ is identified (say, at position i), the set of currently *extendible* subsequences are extended according to the following principles.

- Mark the subsequences that end prior to position $i - 1$ as in Phase C.

- Only the dominating subsequence in Phase B is extended to include the newly discovered match.

- For each subsequence in Phase A, simply extend it by one repetition and check whether the transition point to Phase B is reached. If so, mark this subsequence as in Phase B. If multiple subsequences are in Phase B, then only the dominating one is retained.

- The subsequence with most repetitions in Phase B and C is identified and used to update the *longest valid subsequence* for pattern $(d, *, \ldots, *)$.

- The dominating subsequence in Phase C is also extended. The newly discovered match serves as the beginning of a new segment and the subsequence transits to Phase A.

To address the phase transition issues, the algorithm maintains three separate data structures. The *ongoing_seq* queue tracks all subsequences in Phases A and B. The *valid_seq* queue tracks the potential subsequences in Phase C. The elements in these two queues are overlapped as the transition from Phase B to Phase C is fuzzy. In fact, every subsequence in Phase B will appear in both *ongoing_seq* and *valid_seq* queues. Finally, there is the *longest_seq* that tracks the longest subsequence on a pattern detected so far. We now describe the contents of the various data structures after scanning the ith position of the input sequence.

- *longest_seq*: It contains the longest valid subsequence that is known (at position i) to be not extendible. Since we have no knowledge about the data behavior after position i, only valid subsequences that end prior to position $(i - max\_dis - 1)$ are guaranteed to be not extendible at this moment. We cannot determine the extendibility of any valid subsequence that ends on or after position $(i - max\_dis - 1)$. Therefore, *longest_seq* is the longest valid subsequence that ends prior to position $(i - max\_dis - 1)$.

- *onging_seq*: It contains a set of subsequences that are currently being extended, whose last segment may or may not have enough repetitions to become valid. As we will explain later, the ending position of these subsequences are between i and $(i + l - 1)$ where l is the period length[2]. Thus, we can organize them by their ending positions via a queue structure. Each *entry* in the queue holds a set of subsequences ending at the same position as illustrated in Figure 3.6(a). For example, if we want to verify the pattern $(d_1, *, *)$ against the sequence in Figure 3.6(c) with thresholds $min\_rep = 3$ and $max\_dis = 5$, the *ongoing_seq* queue is illustrated in Figure 3.6(d) after processing the d_1 that occurs in the 12th position. There are three subsequences being extended, one of which (i.e., S_1) ends at position 14 (D_7 is the last match) while the rest (i.e., S_2 and S_3) end at position 13 (D_6 is the last match). We need to maintain both S_2 and S_3 because both of them have a chance to grow to the longest subsequence. Even though S_2 is

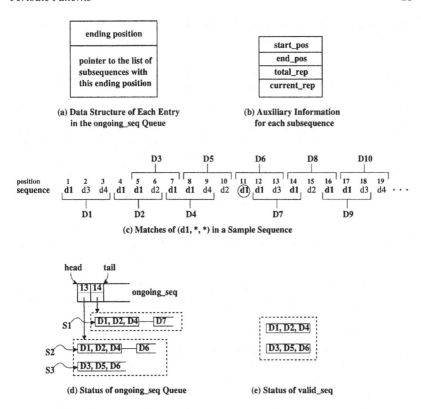

Figure 3.6. *ongoing_seq* and *valid_seq* data structures

longer than S_3, it is not a valid subsequence yet since the last segment does not meet the *min_rep* requirement. Therefore, we cannot discard S_3 at this moment.

- *valid_seq*: It contains a set of valid subsequences that may be extendible. Figure 3.6(e) shows the *valid_seq* set which consists of two valid extendible subsequences. It is necessary to keep them since we may need to extend a valid subsequence multiple times by appending different matches of the pattern. For example, the segment $[D_1, D_2, D_4]$ was extended twice as shown in Figure 3.6(d).

For each subsequence in either *ongoing_seq* or *valid_seq*, we also keep track of the starting position (*start_pos*), the ending position (*end_pos*), the number of overall repetitions (*total_rep*), and the number of repetitions of the last segment (*current_rep*) as shown in Figure 3.6(b) to facilitate the tracking.

The LSI algorithm scans through the input data sequence, for each occurrence of symbol d at position i in sequence D, we have a match from position i to $(i + l - 1)$. Segments of pattern $(d, *, \ldots, *)$ are identified and can be used to extend previously generated subsequences if applicable. The following steps are taken sequentially after a match is detected at position i.

1 The *valid_seq* is first examined to remove all subsequences whose ending position is more than *max_dis* (i.e., the maximum disturbance threshold) away from the current position (i.e., position i). Note that these subsequences cannot be extended any further because of the violation of maximum disturbance requirement. Thus, it is not necessary to keep them in *valid_seq*. At the same time, for each removed subsequence *seq* from *valid_seq*, we compare it with *longest_seq* and update *longest_seq* if necessary.

2 The *ongoing_seq* queue is then investigated. An iteration is taken where each time the entry at the head of the queue is examined until the queue is empty or we reach an entry with ending position on or after i.

   (a) If the ending position of the subsequences in this entry is prior to $(i - 1)$, then the last segment of every subsequence in this entry cannot be extended further. For example, when the circled d_1 in Figure 3.6(c) is reached, we know that the segment $[D_1, D_2, D_4]$ had ended. We can simply dequeue this entry and discard it. The rationale is that we do not have to immediately extend these subsequences by initiating a new segment starting from the current match because all valid extendible subsequences in this entry are already in *valid_seq* and will be examined in Step 3.

   (b) If the ending position is exactly at $(i - 1)$, the last segment of the subsequences in this entry can be extended to include the current match. (The circled d_1 in Figure 3.6(c) also informs us that the segment $[D_3, D_5]$ can be extended to include D_6.) The following steps are taken sequentially.

      i We append the current match to each subsequence in this entry and update the auxiliary data associated with them accordingly. The ending position of these subsequences is also updated to $(i + l - 1)$.

      ii If there are multiple valid subsequences in this entry (i.e., whose *current_rep* satisfies the minimum repetition requirement), then only the subsequence with the largest *total_rep* value is retained, the rest is discarded. It is obvious that all discarded subsequences here are dominated by the retained one. Hence, the discard would not impact the correctness of the algorithm while the efficiency can be improved.

      iii After Step ii, there can be at most one valid subsequence in this entry. If there exists one valid subsequence, then, as a potential Phase C candidate transited from Phase B, this subsequence is replicated to *valid_seq*. (Note that it still remains in *ongoing_seq*.) This gives a valid subsequence the opportunity to be extended in multiple ways concurrently. For example, in Figure 3.6(d), both S_1 and S_2 are extended from the valid segment $[D_1, D_2, D_4]$. It is necessary even if the last segment of this subsequence is still extendible. Because this subsequence may dominate all subsequences in *valid_seq* at some later position and in turn will be extended.

iv Finally, this entry (now ending at position $(i + l - 1)$) is moved from the head of the *ongoing_seq* queue to the tail of the queue.

3 In *valid_seq*, the subsequence *seq* that ends prior to position i and dominates all other subsequences with ending position prior to i is identified[3]. If *seq* does not end at position $(i - 1)$, then it is used to create a new subsequence *new_seq* by extending *seq* to include the current match[4]. The interval between the ending position of *seq* and i is treated as disturbance. *new_seq* is, then, inserted into the entry with ending position $(i+l-1)$ in *ongoing_seq* queue. (If such entry does not exist, a new entry will be created and added to the tail of the *onging_seq* queue.) This signals the transition of the subsequence from Phase C to Phase A.

After the entire sequence is scanned, the subsequence which has the largest *total_rep* value in *valid_seq* $\cup$ {*longest_seq*} is returned.

Example

Figure 3.7(a) shows a sequence of symbols which is the same as in Figure 3.6(c) where the status of the various data structures after processing the 7th occurrence of d_1 at position 12 are shown in Figure 3.6(d) and (e) (The *longest_seq* is still empty.). The process of the 8th, 9th, and 10th occurrences of d_1 is illustrated in Figure 3.8 while the change to the data structures at each step is shown in Figure 3.7(b), (c), and (d).

From the above example, we can make the following observations.

- For *ongoing_seq*, there can be at most l entries, each of which corresponds to a different ending position between i and $(i+l-1)$. Furthermore, in each of these entries, each subsequence has a distinct length of the last segment (i.e., a distinct *current_rep*). Also, there can be only one subsequence with a *current_rep* larger than or equal to *min_rep* due to Step 2(b)ii.

- For *valid_seq*, each element has a distinct ending position between $(i - max_dis - 1)$ and $(i + l - 1)$ due to Step 1.

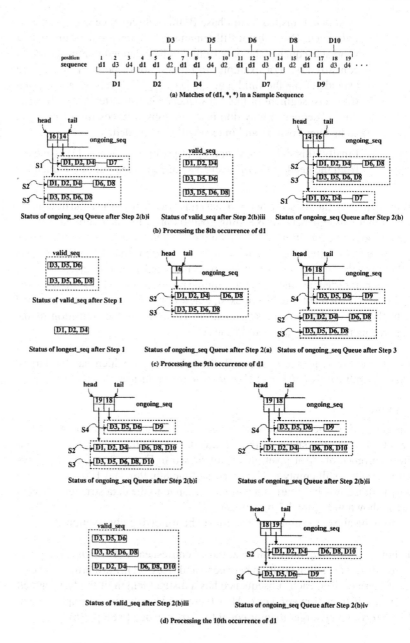

Figure 3.7. Status of *ongoing_seq*, *valid_seq*, and *longest_seq*

Step	8th occurrence of d1 at position 14	9th occurrence of d1 at position 16	10th occurrence of d1 at position 17
1	Both subsequences in valid_seq end less than 5 positions away from the current position (i.e., position 14). As a result, no subsequence will be discarded from valid_seq.	$\boxed{\text{D1, D2, D4}}$ ends at position 9 which is more than 5 positions away from the current position. Thus, it is removed from valid_seq since we can not get any new extension from it. It is then put in longest_seq. (longest_seq was empty previously.)	valid_seq remains unchanged since both subsequences end less than 5 positions away from the current one.
2(a)	This step is skipped since no entry in onging_seq ends before position 13.	The entry with ending position 14 is removed from the onging_seq.	This step is skipped since no entry in onging_seq ends before position 16.
2(b)i	All subsequences in the entry with ending position 13 are extended to include D8. The ending position is also updated to 16 to reflect the change.	This step is skipped since no entry in onging_seq ends at position 15.	All subsequences in the entry with ending position 16 are extended to include D10. The ending position is also updated to 19 to reflect the change.
2(b)ii	This step is skipped since only one valid subsequence (i.e., S3) exists in this entry.		S3 is removed since S2 is also valid and longer than S3.
2(b)iii	$\boxed{\text{D3, D5, D6, D8}}$ is replicated to valid_seq.		$\boxed{\text{D1, D2, D4}}$—$\boxed{\text{D6, D8, D10}}$ is replicated to valid_seq.
2(b)iv	This entry (which ends at position 16) is moved from the head of ongoing_seq to the tail of it.		This entry (which ends at position 19) is moved from the head of ongoing_seq to the tail of it.
3	Since $\boxed{\text{D3, D5, D6}}$ ends at position 13, no new subsequence needs to be generated.	S4 is generated by extending $\boxed{\text{D3, D5, D6}}$ to include D9. A new entry is created with ending position 18 and is appended at the tail of ongoing_seq.	Since $\boxed{\text{D3, D5, D6, D8}}$ ends at position 16, no new subsequence needs to be generated.

Figure 3.8. Example for Illustration

Complexity Analysis

We first analyze the time complexity, then the space complexity of the LSI algorithm.

Time Complexity

The sizes of $valid\_seq$ and $ongoing\_seq$ are essential for analyzing the time complexity of the LSI algorithm. After processing one target symbol d at position i, at most one valid subsequence will be inserted to $valid\_seq$ (in Step 2(b)iii with ending position $(i + l - 1)$ as indicated in Step 2(b)i). It follows that every subsequence in $valid\_seq$ has different ending positions. In addition, after processing each match (starting at position i), the ending position of all subsequences in $valid\_seq$ is between $(i - max\_dis - 1)$ and $(i + l - 1)$ (inclusively). As a result, there are at most $max\_dis + l + 1$ subsequences in $valid\_seq$. Thus, the complexity of Step 1 is $O(max\_dis + l)$. Since it is invoked once for each match of the pattern, the overall time complexity of this step for processing LSI for a given symbol d is $O\left(n_d \times (max\_dis + l)\right)$ where n_d is the number of occurrences of d.

Within the entire procedure, each entry removed from the head of $ongoing\_seq$ in Step 2(a) is associated with a distinct ending position of the target pattern. Since there are at most n_d distinct ending positions (each of which corresponds to a match) in Step 2(b)i, at most n_d entries are ever removed from $ongoing\_seq$

for a given symbol d and a given period. Step 2(a) can be invoked at most n_d times. Therefore, the overall complexity for Step 2(a) is $O(n_d)$.

Consider the course of an entry r in *ongoing_seq* from the time it is first initialized in Step 3 to the time it is permanently discarded in Step 2(a). It is easy to show by induction that each subsequence in r has a distinct value of *current_rep*. This claim holds trivially when r is initialized in Step 3 where only one subsequence is in r. At each subsequent time a new subsequence *new_seq* is added to r (in Step 3) as a result of processing a match M, the value of *current_rep* of *new_seq* is always 1 since M is the only match in the last segment of *new_seq* (e.g., S_2 in Figure 3.6(d)). In contrast, the *current_rep* of other subsequences in r are at least 2 after their last segments were extended to include M (e.g., S_3 in Figure 3.6(d)). Therefore, *new_seq* has a different value of *current_rep* from other subsequences in r. Thus, we can conclude that each subsequence in r holds a distinct value of *current_rep*. Since there is at most one subsequence whose last segment has at least *min_rep* repetitions (due to Step 2(b)ii), the number of subsequences in r is bounded by *min_rep*. The complexity of each invocation of Step 2(b) is $O(min\_rep)$. At any time, each entry in *ongoing_seq* is associated with a distinct ending position. When processing a match starting at position i, at most one entry has ending position $i - 1$. The overall complexity of Step 2(b) for a given symbol d and a given period l is $O(n_d \times min\_rep)$.

As we explained before, at most $(max\_dis + l + 1)$ subsequences are in *valid_seq* at any time. In turn, it takes $O(max\_dis + l)$ time complexity each time Step 3 is invoked. This brings to a total complexity of $O(n_d \times (max\_dis + l))$. In summary, the overall complexity of the LSI algorithm for a given symbol and period length l is

$$O\left(n_d \times (min\_rep + max\_dis + l)\right).$$

For a given period length l, the complexity to find the "longest" subsequence for all symbols is hence

$$O\left(\sum_{\forall d} n_d \times (min\_rep + max\_dis + l)\right)$$

which is $O\left(N \times (min\_rep + max\_dis + l)\right)$ where N is the length of the input sequence. Thus, the time complexity to discover the "longest" single-symbol subsequence for all periods and symbols is

$$O\left(N \times L_{max} \times (min\_rep + max\_dis + L_{max})\right)$$

where L_{max} is the maximum period length. This is the worst case complexity. Since the distance-based pruning may prune a large number of symbol and period pairs, the real running time could be much faster. In addition, we will propose several techniques that can reduce the time complexity of the LSI algorithm later..

Space Complexity

There are two main data structures in LSI, *ongoing_seq* and *valid_seq*. For each symbol d and a given period length l, the size of *valid_seq* is $O(max\_dis + l)$ whereas the size of *ongoing_seq* is bounded by n_d since for each occurrence of d, one new subsequence is inserted to *ongoing_seq*. Furthermore, since each entry of *ongoing_seq* has at most *min_rep* subsequences, the size of *ongoing_seq* is $O(min(n_d, min\_rep \times l))$. Therefore, the space complexity to find the "longest" subsequences for all symbols and a given period length l is $O(S + T)$ where

$$S = (max\_dis + l) \times Num\_Symbols,$$

$$T = min(N, min\_rep \times l \times Num\_Symbols,$$

and $Num\_Symbols$ is the number of symbols in the input sequence. The overall space complexity for all possible period lengths is $O(U + V)$ where

$$U = (max\_dis + L_{max}) \times Num\_Symbols \times L_{max}$$

$$V = min(N \times L_{max}, min\_rep \times L_{max}^2 \times Num\_Symbols).$$

The above space complexity analysis is again a theoretical bound in the worst case; however, the space requirement is much smaller in practice shown in empirical studies. Thus, in reality, all data structures can be easily fit into main memory.

Improvement of the LSI Algorithm

The time complexity of the LSI algorithm can be improved further. One way is to use a queue to store *valid_seq* and a heap to index all subsequences in *valid_seq* according to their *total_rep*. Each time a subsequence is inserted into *valid_seq*, it is added to the end of the queue. This would naturally make all subsequences lie in the queue in ascending order of their ending positions. Thus, Step 1 can be easily accomplished by dequeue obsolete subsequence(s) from the head of the queue. Of course, each of such operation would incur $O(\log(max\_dis + l))$ overhead to maintain the indexing heap. However, in virtue of the heap indexing, each invocation of Step 3 only requires $O(\log(max\_dis + l))$ time complexity for period length l. Therefore, the overall complexity of LSI algorithm for all period lengths and symbols is reduced to

$$O\left(N \times L_{max} \times (min\_rep + \log(max\_dis + L_{max}))\right). \tag{3.1}$$

Proof of Correctness

LEMMA 3.1 *The last segment of any invalid subsequence removed from the data structure ongoing_seq is not extendible.*

Proof. Step 2(a) is the only place we may remove an invalid subsequence from *ongoing_seq*. Assume that the subsequence ends at position k ($k < i - 1$). It must be true that no match starts on position $k + 1$[5]. Thus, the last segment of the invalid subsequence is not extendible. □

LEMMA 3.2 *At least one prefix of each longest valid subsequence has been put in both ongoing_seq and valid_seq.*

Proof. Consider the starting position, say j, of a longest valid subsequence X. All valid segments starting before position ($j - min\_rep \times l$) have to end before position ($j - max\_dis - 1$). (Otherwise, a longer valid subsequence can be formed by extending X "backwards" to include additional valid segment(s). This contradicts the assumption that X is the longest valid subsequence.) As a result, *valid* is empty at position j. Then a new subsequence (denoted by Y) starting at position j consisting of one match of the pattern is added to *ongoing_seq*. In addition, j is the starting position of a valid segment (because it is the starting position of X). By Lemma 3.1, Y will stay in *ongoing_seq* until it grows to become a valid subsequence (i.e., cumulates at least min_rep repetitions). When Y is extended to a valid subsequence (denoted by Z), Z will be replicated to *valid_seq* because Z is the longest one in *ongoing_seq*. (All other subsequences in *ongoing_seq* start later than Z.) Thus, this lemma holds. □

LEMMA 3.3 *After processing each match, all valid subsequences that are in ongoing_seq also appear in valid_seq.*

Proof. All valid subsequences in *ongoing_seq* are generated in Step 2(b)i (some of which might be removed immediately in Step 2(b)ii). The remaining one (if applicable) is then replicated in *valid_seq* (Step 2(b)iii). □

By Lemmas 3.1 and 3.2, for any longest valid subsequence, either it is fully generated and used to update *longest_seq* or one of its valid prefix is removed from *valid_seq* without being extended further. Now, consider the processing of a match M starting on position i.

LEMMA 3.4 *After processing a match M that starts at position i, one of the longest valid subsequences that end between position ($i - max\_dis - 1$) and ($i + l - 1$) is in valid_seq.*

Proof. When processing a match M at positions i to ($i + l - 1$), a valid subsequence X with ending position k ($i - max\_dis - 1 \leq k \leq i - 1$) may not be extended to include M due to one of the following two reasons.

1 $k = i - 1$. X was removed from *ongoing_seq* in Step 2(b)ii because of the existence of another valid subsequence Y ending at position k in *ongoing_seq* such that Y dominates X. Y is chosen to be extended to include M and to be retained in *ongoing_seq* for potential further extension.

2 Otherwise, X is in *valid_seq* and is dominated by some other valid subsequence Y in *valid_seq*, which is extended to include M and added into *ongoing_seq* (Step 3).

In summary, the only reason to stop extending X is that X is dominated by some other valid subsequence Y that is extended to include M and resides in *ongoing_seq*. By Lemma 3.3, all valid subsequences in *ongoing_seq* is in *valid_seq*. Therefore, after processing each match, any valid subsequence ending between position $(i - max\_dis - 1)$ and $(i + l - 1)$ is either itself in *valid_seq* or is dominated by some other valid subsequence in *valid_seq*. In other words, at least one of the longest valid subsequences that end between position $(i - max\_dis - 1)$ and $(i + l - 1)$ is in *valid_seq*. □

LEMMA 3.5 *After processing a match M that starts at position i, one of the longest valid subsequences that end prior to position $(i - max\_dis - 1)$ is in longest_seq.*

Proof. This proof is trivial because every time a subsequence is removed from *valid_seq* (due to an obsolete ending position), the *longest_seq* is updated if necessary. □

The following theorem is a direct inference of Lemma 3.4 and 3.5.

THEOREM 3.6 *After processing the entire sequence, longest_seq holds one of the longest valid subsequences.*

Complex Patterns

After discovering the single patterns, valid subsequences of different symbols may be combined to form a valid subsequence of multiple symbols of the same period. We employ a level-wise search algorithm, which generates the subsequences of i-patterns based on valid subsequences of $(i - 1)$-patterns with the same period length. To efficiently prune the search space, we use two properties: one is the **symbol** property, and the other is the **segment** property.

PROPERTY 1.1 (**Symbol property**) *If a pattern P is valid, then all of its generalizations are valid.*

PROPERTY 1.2 (**Segment property**) *If $D' = d_j, d_{j+1}, d_{j+2}, \ldots, d_k$ is a valid segment for pattern P, then D' is also a valid segment of all generalizations of P.*

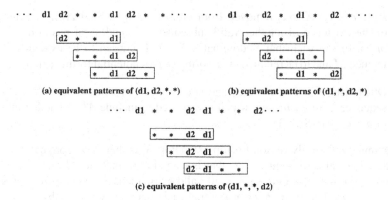

Figure 3.9. Equivalent Complex Patterns

Since these two properties are straightforward, we would omit the proof. Based on these properties, we can prune the candidates of a valid pattern efficiently. For example, if two patterns $(d_1, *, *, *)$ and $(d_2, *, *, *)$ are valid, then three candidate 2-patterns can be generated: $(d_1, d_2, *, *)$, $(d_1, *, d_2, *)$ and $(d_1, *, *, d_2)$. As shown in Figure 3.9, all other 2-patterns of period 4 containing d_1 and d_2 are equivalent to one of these three patterns with a shifted starting position. Similarly, $(d_1, d_2, d_3, *)$ can become a candidate 3-pattern only if $(d_1, d_2, *, *)$, $(d_1, *, d_3, *)$, and $(d_2, d_3, *, *)$ are all valid.

After the candidate set of valid i-patterns is generated, then a similar algorithm to LSI is executed to verify whether these candidates are indeed valid patterns. As a side product, the LSI algorithm also finds the valid subsequence with the most repetitions for each valid pattern.

Discussion

Parameters Specification

In our approach, the mining results can be effected by the choice of the two parameters min_rep and max_dis. When the parameters are not set properly, noises may be qualified as patterns. We use an iterative method to set the proper value for these two parameters. After discovering patterns for a given pair of min_rep and max_dis, we prune those discovered patterns according to the statistical significance. For example, if $(a, *, *)$ is a discovered pattern and the expected continuous repetition of this pattern in a random sequence exceeds the min_rep threshold, then we can conclude that this pattern may occur due to chance and it should be discarded. Notice that there may not be a uniform threshold of min_rep and max_dis for statistical significance since the probability of occurrence of two patterns may be difference. For instance, the occurrence probability of $(a, *, *)$ should be higher than that of $(a, b, *)$. After pruning, if the number of remaining patterns is too small, and we can adjust the parameters of min_rep and max_dis, e.g., reducing min_rep or increasing max_dis, and mine patterns again. This process terminates when there is a sufficient number of patterns discovered.

Noises

There may be many types of noises in real applications. The parameter max_dist is employed to recognize the noises between segments of perfect repetitions of a pattern. There may exist other types of noises, e.g., intra-pattern noise. Let $(a, c, *, b, *, *, b)$ be a pattern. The segment $accbaaab$ may be the occurrence of this pattern with some noise. (There is an extra symbol between the last two bs.) We can modify the definition of asynchronous pattern slightly to recognize this type of noises. If a segment is very similar to a pattern (within a certain degree), then we consider the segment as a repetition of the pattern. Without major modification, our algorithm should be able to handle this type of noise.

Extensions

In some application, e.g., sensor network, multiple events may occur simultaneously. As a result, it is possible that multiple symbols may occur in the same position within a sequence. The proposed approach can be easily extended to handle this situation. We only need to modify one step in our algorithm. When generating candidates, we need to take into account all possible subsets of symbols at a time slot. For example, if symbols A, B occurred at the same time slot, then during candidate generation phase, we need to consider four possible candidates for this time slot, $\{A\}, \{B\}, \{A, B\}$, and $\{*\}$.

The other possible extension of the asynchronous pattern is to discover possible sequential rules, e.g., "A is followed by B with a chance 50% in a given subsequence". To find this type of sequential rules, the minimum repetitions can be viewed as support. The asynchronous patterns discovered by our algorithm can be considered as the patterns that satisfy the support threshold. In the post-process step, we can verify whether the rules satisfy the confidence requirement. To quality the rule "A is followed by B with a chance 50%, then by C with a probability 75% in a given subsequence", all three patterns $(a, *, \dots)$, $(b, *, \dots)$, and $(c, *, \dots)$ have to be valid for the a sufficient long portion of the sequence. Next, we can verify whether the confidence requirement (e.g., 50% and 75%) is also satisfied.

Experimental Results for Asynchronous Patterns

We implemented the PatternMiner in C programming language and it is executed on an IBM RS-6000 (300 MHz CPU) with 128MB running an AIX operating system.

Real Data Set

We first apply our model to a real trace of a web access log. Scour is a web search engine specialized in multimedia content search whose URL is "http://www.scour.net". Since early of 2000, the average daily number of hits on Scour has grown over one million. A trace of all hits on Scour between March 1 and June 8 (total 100 days) were collected. The total number of accesses is over 170 million. Then the entire trace is summarized into a sequence as follows. The trace is divided into 10 minute intervals. The number of hits during each 10 minute interval is calculated. Finally, we label each interval with a symbol. For example, if the number of hits is between 0 and 4999, then this interval is labeled as a, if the number of hits is between 5000 and 9999, then this interval is labeled as b, and so on. The summarized sequence consists of 14400 occurrences of 71 distinct symbols.

There exist some interesting patterns discovered by our algorithm. When min_rep and max_dis are set to 4 and 200, respectively, there are overall 212 patterns discovered. The following is some example of discovered patterns. There exists a pattern $(b,b,\ b)$ in weekdays between 3am and 8:30am EST. Another pattern $(c,\ c,\ c)$ occurs during 11am to 5pm EST weekday.

In the above experiment, the overall length of the sequence is relatively short (14400), hence, all three mining processes are done in less than 1 minute. To further understand the behavior of our proposed asynchronous pattern mining algorithm, we constructed four long synthetic sequences and the sensitive analysis of our algorithm on these sequences is presented in the following section.

Synthetic Sequence Generation

For the purpose of evaluation of the performance of the PatternMiner, we use four synthetically generated sequences. Each sequence consists of 1024 distinct symbols and 20M occurrences of symbols. The synthetic sequence is generated as follows. First, at the beginning of the sequence, the period length l of the next pattern is selected based on a geometric distribution with mean μ_l. The number of symbols involved in a pattern is randomly chosen between 1 and the period l. The number of valid segments is chosen according to a geometrical distribution with mean μ_s. The number of repetitions of each valid segment follows a geometrical distribution with mean μ_r. After each valid segment, the length of the disturbance is determined based on a geometrical distribution with mean μ_d. This process repeats until the length of the sequence reaches 20M.

Four sequences are generated based on values of μ_l, μ_s, μ_r, and μ_d in Table 3.1. In the following experiments, we always set $L_{max} = 1000$ and for each period l, $max\_dis = \frac{min\_rep \times l}{4}$.

Data Set	μ_l	μ_s	μ_r	μ_d
$DS1$	5	5	50	50
$DS2$	5	5	1000	1000
$DS3$	100	1000	50	50
$DS4$	100	1000	1000	1000

Table 3.1. Parameters of Synthetic Data Sets

Distance-based Pruning

In this section, we are investigating the effects of distance-based pruning. Without the distance-based pruning, there could be as many as $\mid \Im \mid \times L_{max}$ different single patterns $(d, *, \ldots, *)$ $(\forall d \in \Im)$. Figure 3.10(a) shows the fraction of patterns eliminated by distance-based pruning. It is evident that the number of pruned patterns is a monotonically increasing function of the min_rep threshold. With a reasonable min_rep threshold, only a small portion of potential single patterns need to be validated and used to generate candidate complex pattern.

Figure 3.10(b) shows the space utilized by $ongoing\_seq$ and $valid\_seq$ for all patterns in the LSI algorithm. The real space utilization is far less than the theoretical bound shown above due to the following two reasons. First a large number of candidate patterns are pruned. (More than 90% as shown in Figure 3.10(a).) Secondly, in the theoretical analysis, we consider the worst case for the space utilization of $ongoing\_seq$, however, in reality, the space occupied $ongoing\_seq$ is much less than the theoretical bound.

Pattern Verification

After the distance-based pruning, each remaining potential single pattern is validated through the LSI algorithm. Since the validation process (even for a given pattern) is not a trivial task, in this subsection, we demonstrate the efficiency of our LSI algorithm by comparing it with a reasonable two stage (TS) algorithm. In the TS algorithm, for a given pattern, all valid segments are first discovered, then all possible combinations of valid segments are tested and the one with the most repetition is chosen. Figure 3.11 shows the average elapse time of validating a pattern. (Note that the Y-axis is in log scale.) It is evident that LSI can outperform the TS algorithm by at least one order of magnitude regardless the min_rep threshold.

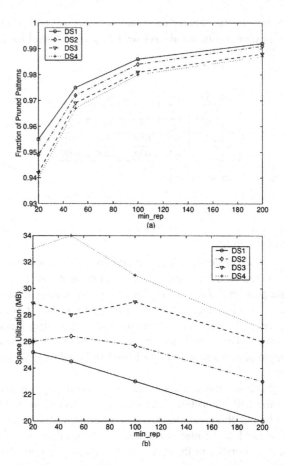

Figure 3.10.   Effects of Distance-based Pruning

Overall Response Time

Figure 3.12 shows the overall response time of PatternMiner to find all patterns. The x-axis shows the value of min_rep whereas the y-axis shows the overall response time of PatternMiner. The higher the min_rep threshold, the shorter the overall response time. This is in contrast to Formula 3.1 due to the effect of distance-based pruning shown in Figure 3.10.

2.   Meta-periodic Patterns

In addition, due to the changes of system behavior, some pattern may be only notable within a portion of the entire data sequence; different patterns may present at different places and be of different durations. The evolution among

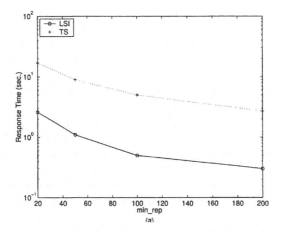

Figure 3.11.   LSI Algorithm vs. TS Algorithm

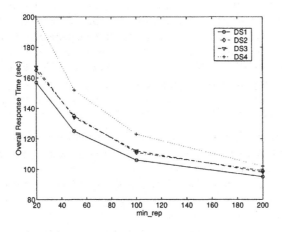

Figure 3.12.   Overall Response Time of PatternMiner

patterns may also follow some regularity. Such regularity, if any, would be of great value in understanding the nature of the system and building prediction models. Consider the application of *inventory replenishment*. The history of inventory refill orders can be regarded as a symbol sequence. For brevity, let's only consider the replenishment of flu medicine. Figure 3.13(a) shows the history of refill orders of a pharmacy during 1999 and 2000 on a weekly basis. The symbol "r" means a refill order of flu medicine was placed in the corresponding week while "-" represents that no flu medicine replenishment was made in that week. It is easy to see that the replenishment follows a biweekly pattern during the first half of each year and a triweekly cycle during the second half of each year. This seasonal fluctuation also forms a high level periodic pattern (The period length is one year). However, such high level patterns may not be expressible by any previous model (defined in terms of raw symbols) due to the presence of noise, even when the noise is very limited. In the above example, a major outbreak of flu caused a provisional replenishment at the 4th week of 1999 (Figure 3.13(a)). Afterwards, even though the replenishment frequency is back to once every other week, the occurrences of all subsequent replenishments become misaligned. Even though the biweekly replenishment cycle was notable in the first two quarters of both 1999 and 2000, the corresponding portions in the data sequence have a different layout of replenishment occurrences. This characteristic determines that the representation of the above two-level periodicity is beyond the expressive power of any traditional model of periodic patterns that only takes raw symbols as components. In any traditional model, each symbol specified in a pattern uniquely matches its counterpart in the data sequence, and all occurrences of a pattern have to share a unique common layout. This inherent limitation would prevent many interesting high level patterns from being captured. Note that, even if the period length (i.e., 52 weeks) is given[6], the only annual pattern that is able to be generated under the traditional model via pairwise comparison of symbols corresponding to each week is shown in Figure 3.13(b). The symbol "*" denotes the don't care position[7] and can match any symbol on the corresponding position. Clearly, little information is conveyed in this pattern as the important two-level periodicity is completely concealed.

To tackle the problem, we propose a so-called **meta-pattern** model to capture high level periodicities [8]. A meta-pattern may take occurrences of patterns/meta-patterns (of lower granularity) as components. In contrast, we refer to the patterns that contain only raw symbol(s) as the **basic patterns**, which may be viewed as special cases of meta-patterns. In general, the noise could occur anywhere, be of various duration, and even occur multiple times within the portion where a pattern is notable as long as the noise is below some threshold. Even though the allowance of noise plays a positive role in characterizing system behavior in a noisy environment, it prevents such a meta-pattern from being represented in the form of an (equivalent) basic pattern. The model of meta-pattern provides a more powerful means to periodicity representation. The recursive nature of meta-pattern not only can tolerate a greater degree of noises/distortion, but also can capture the (hidden) hierarchies of pattern evolutions, which may not be expressible by previous models. In the previous example, the biweekly and the triweekly replenishment cycles can be easily represented by $P_1 = (r : [1,1], * : [2,2])$ and $P_2 = (r : [1,1], * : [2,3])$, respectively, where the numbers in the brackets indicate the offset of the component within the pattern. The two-level periodicity can be easily represented

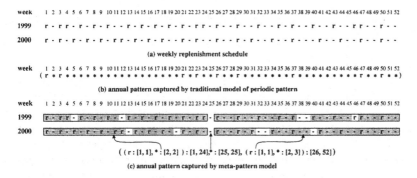

Figure 3.13. Meta Pattern

as $(P_1 : [1, 24], * : [25, 25], P_2 : [26, 52])$ which can be interpreted as the pattern P_1 repeats at the first 24 weeks and the pattern P_2 repeats from week 26 to week 52. As shown in Figure 3.13 (c), each rectangle box denotes the portion where the corresponding low level pattern (i.e., either $(r : [1, 1], * : [2, 2])$ or $(r : [1, 1], * : [2, 3]))$ is notable.

Unfortunately, the flexibility of meta-pattern poses serious challenges in the discovery process, which may not be encountered in mining basic patterns.

- While a basic pattern has two degrees of freedom: the period (i.e., the number of positions in the pattern) and the choice of symbol for each single position, a meta-pattern has an additional degree of freedom: the length of each component in the pattern. It is incurred by the fact that a component may occupy multiple positions. This extra degree of freedom increases the number of potential meta-pattern candidates dramatically.

- Many patterns/meta-patterns may collocate or overlap for any given portion of a sequence. As a result, during the meta-pattern mining process, there could be a large number of candidates for each component of a (higher level) meta-pattern. This also aggravates the mining complexities.

Therefore, how to identify the "proper" candidate meta-patterns is crucial to the overall efficiency of the mining process, and will be the focus of the algorithmic part of the section. To tackle this problem, we employ a so called *component location property*, in addition to the traditionally used Apriori property, to prune the search space. This is inspired by the observation that a pattern may participate in a meta-pattern only if its notable portions exhibit a certain cyclic behavior. A *computation-based* algorithm is devised to identify the potential period of a meta-pattern and for each candidate period, the potential components and their lengths within the meta-pattern. The set of all meta-patterns can be categorized according to their structures and are evaluated in a designed order so that the pruning power provided by both properties can be fully utilized.

Model of Meta-Patterns

Let $\Im = \{a, b, c, \dots\}$ be a set of literals. A traditional periodic pattern [4, 9] (e.g., asynchronous pattern) consists of a tuple of k components, each of which is either a literal or "*". k is usually referred to as the **period** of the pattern. "*" can be substituted for any literal and is used to enable the representation of partial periodicity. For example, $\dots, a, b, c, a, b, d, a, b, b, \dots$ is a sequence of literals and $(a, b, *)$ [8] represents that the incident "b following a" occurs for every 3 time instances in the data sequence. The period of this pattern is 3 by definition. Note that the third component in the pattern is filled by a "*" since there is no strong periodicity presents in the data sequence with respect to this component. Because a pattern may start anywhere in the data sequence, only patterns whose first component is a literal in $\Im$ need to be considered. In this chapter, we refer to this type of patterns as **basic** patterns as each component in the pattern is restricted to be either a literal or a "*". In contrast, a **meta-pattern** may have pattern(s)/meta-pattern(s) as its component(s). This enables us to represent complicated basic patterns in a more concise way and to possibly reveal some hidden patterns among discovered ones. Formally, a **meta-pattern** is a tuple consisting of k components $(x_1, x_2, \dots, x_k)$ where each x_i $(1 \le i \le k)$ can be one of the following choices augmented by the offsets of the starting and ending positions of the component with respect to the beginning of the meta-pattern.

- a symbol in $\Im$;

- "don't care" *;

- a pattern/meta-pattern.

We also require that at least one position of a meta-pattern has to correspond to a non "*" component to ensure a non-trivial pattern. For example, $((r : [1, 1], * : [2, 2]) : [1, 24], * : [25, 25], (r : [1, 1], * : [2, 3]) : [26, 52])$ is a meta-pattern with three components: $(r : [1, 1], * : [2, 2])$, *, and $(r : [1, 1], * : [2, 3])$. The **length** of a component is the number of positions that the component occupies in the meta-pattern. In the previous example, the component length of $(r : [1, 1], * : [2, 2])$ is 24. We also say that 52 is the **span** of this meta-pattern, which is equal to the sum of the length of all components in the meta-pattern. This pattern can be interpreted as *"the pattern $(r : [1, 1], * : [2, 2])$ is true for 24 positions (or weeks in previous example) followed by the pattern $(r : [1, 1], * : [2, 3])$ for 27 positions with a gap of one position in between, and such a behavior repeats for every 52 positions"*. For brevity, we sometimes omit the augmenting offset of a component if the length of the component is only one position. For example, $(r, *)$ is the abbreviation of $(r : [1, 1], * : [2, 2])$

and $((r, *) : [1, 24], *, (r, * : [2, 3]) : [26, 52])$ is equivalent to $((r : [1, 1], * : [2, 2]) : [1, 24], * : [25, 25], (r : [1, 1], * : [2, 3]) : [26, 52])$. It is obvious that the meta-pattern is a more flexible model than the basic pattern and the basic pattern can be viewed as a special (and simpler) case of the meta-pattern. Because of the hierarchical nature of the meta-pattern, the concept of *level* is introduced to represent the "depth" of a meta-pattern. By setting the level of basic pattern to be 1, the **level** of a meta-pattern is defined as the maximum level of its components plus 1. According to this definition, the level of $(r, * : [2, 3])$ is 1 and the level of $P_1 = ((r, *) : [1, 24], *, (r, * : [2, 3]) : [26, 52])$ is 2. Note that the components of a meta-pattern do not have to be of the same level. For instance, $(P_1 : [1, 260], * : [261, 300])$ is a meta-pattern (of level 3) which has a level-2 component and a level-1 component.

All terminologies associated with the asynchronous patterns presented before (i.e., level-1 patterns) can be generalized to the case of meta-patterns (i.e., higher level patterns). We call the level-1 patterns the basic patterns. We now give a brief overview of terms defined for basic patterns. Given a symbol sequence $D' = d_1, d_2, \ldots, d_s$ and a basic pattern $P = (p_1, p_2, \ldots, p_s)$, we say D' **supports** P iff, for each $i(1 \leq i \leq s)$, either $p_i = *$ or $p_i = d_i$. D' is also called a **match** of P. Given a pattern P and a symbol sequence D, a list of j disjoint matches of P in D is called a **segment** with respect to P iff they form a contiguous portion of D. j is referred to as the **number of repetitions** of this segment. Such a segment is said to be a **valid segment** iff j is greater than or equal to the required minimum repetition threshold min_rep. A **valid subsequence** in D (with respect to P) is a set of disjoint valid segments where the distance between any two consecutive valid segments does not exceed the required maximum disturbance threshold max_dis. P is said to be a **valid pattern** in D if there exists a valid subsequence in D with respect to P. The parameters min_rep and max_dis, in essence, define the significance of the periodicity and the boundary to separate noise and change of system behavior. The appropriate values of min_rep and max_dis are application dependent and are specified by the user.

Similarly, given a symbol sequence $D' = d_1, d_2, \ldots, d_s$, for any meta-pattern $X = (x_1 : [1, t_1], x_2 : [t_1 + 1, t_2], \ldots, x_l : [t_{l-1} + 1, s])$, D' **supports** X iff, for each component x_i, either (1) x_i is "*" or (2) x_i is a symbol and $d_{t_{i-1}+1} = \cdots = d_{t_i} = x_i$ or (3) x_i is a (meta-)pattern and $d_{t_{i-1}+1}, \ldots, d_{t_i}$ is a valid subsequence with respect to x_i. D' is in turn called a **match** of P. We can define *segment*, *subsequence*, and *validation* in a similar manner to that of a basic pattern. Given a meta-pattern X and a symbol sequence D, a list of j disjoint matches of X in D is called a **segment** with respect to X iff they form a contiguous portion of D. j is referred to as the **number of repetitions** of this segment. Such segment is said to be a **valid segment** iff j is greater than or equal to the required minimum repetitions min_rep. A **valid subsequence** in D (with respect to X) is a set of disjoint valid segments where the distance between any two consecutive valid segments does not exceed the required maximum disturbance max_dis. P is said to be a **valid pattern** in D if there exists a valid subsequence in D with respect to X. Look back to the medicine replenishment example, the max_dis parameter solves the shift problem. Our model can tolerate that the pattern shifts at most max_dis symbols. In other words, our model can capture the patterns that have at least min_rep perfect repetitions continuously and at most max_dis interruption between two portions of perfect repetitions.

In this section, given a symbol sequence and two parameters min_rep and max_dis, we aim at mining valid meta-patterns together with their longest valid subsequences (i.e., the valid subsequence which has the most overall repetitions of the corresponding meta-pattern). Since a meta-pattern can start anywhere in a sequence, we only need to consider those starting with a non "*" component.

Algorithm for Mining Meta-Patterns

The great flexibility of the model poses considerable difficulties to the generation of candidate meta-patterns. Therefore, we will focus on the efficient candidate generation of meta-patterns in the remainder of this chapter. The well-known Apriori property holds on the set of meta-patterns of the same span, which can be stated as follows: *for any valid meta-pattern* $P = (P_1 : [1, t_1], P_2 : [t_1 + 1, t_2], \ldots, P_s : [t_{s-1} + 1, t_s])$, *the meta-pattern constructed by replacing any component* P_i *with "*" in* P *is also valid.* For example, let $X_1 = ((a, b, *) : [1, 19], * : [20, 21])$ and $X_2 = ((a, b, *) : [1, 19], * : [20, 21], (b, c) : [22, 27], * : [28, 30], X_1 : [31, 150])$. If X_2 is a valid meta-pattern, then the pattern $X_3 = ((a, b, *) : [1, 19], * : [20, 21], (b, c) : [22, 27], * : [28, 150])$ (generated by replacing X_1 with "*") must be valid as well. Note that X_2 is a level-3 meta-pattern which has three non "*" components: $(a, b, *)$, (b, c), and X_1; whereas X_3 is a level-2 meta-pattern that has two non "*" components: $(a, b, *)$ and (b, c). Intuitively, X_3 should be examined before X_2 so that the result can be used to prune the search space.

Nevertheless, because of the hierarchical characteristic of the meta-pattern, the Apriori property does not render sufficient pruning power as we proceed to high level patterns from discovered low level patterns. After identifying valid meta-patterns of level l, the brute force method (powered by the Apriori property) to mine patterns of level $l + 1$ is to first generate all possible candidates of level $l + 1$ by taking valid lower level patterns as component(s); and then, verify them against the symbol sequence. While the verification of a base pattern can be performed efficiently (e.g., in linear time with respect to the length of the symbol sequence [9]), the verification for a candidate meta-pattern may be a cumbersome process because of the typically complicated structure of the candidate meta-pattern. In fact, considerable difficulty lies on determining whether a certain portion of the raw symbol sequence corresponds to a valid subsequence of a component of the candidate pattern, especially when the component itself is also meta-pattern. One strategy to speed up the process is to store all valid subsequences of each valid low level pattern when the pattern is verified. Then the procedure of determining whether a portion of the sequence is a valid subsequence of a given component can be accomplished via table look-up operations. Even though this strategy requires additional storage space, it can usually lead to at least an order of magnitude of performance improvement. We will refer this method as the **match-based approach** in the remainder of this chapter. However, this match-based approach is still very cumbersome, and more specifically, suffers from two major drawbacks.

- The number of candidate patterns of a certain level (say level l) is typically an exponential function of the number of discovered lower level meta-patterns. While a basic pattern has two degrees of freedom: the period and the choice of symbol at each position/component, a meta-pattern has an additional degree of freedom: the length of each component. This additional degree of freedom dramatically increases the number of candidate patterns generated. If there are v valid lower level patterns, the number of candidate patterns of span s and with exactly k components for level l is in the order of $\Theta(v^k \times (2k)^s)$.

- There are typically a huge number of valid subsequences associated with each valid pattern even though only a few of them may eventually be relevant. Generating and storing all of them would consume a significant amount of computing resources and storage space, which in turn leads to unnecessary inefficiency.

To overcome these drawbacks, we made the following observation.

PROPERTY 2.1 **(Component Location Property)** *A valid low level meta-pattern may serve as a component of a higher level meta-pattern only if its presence in the symbol sequence exhibits some cyclic behavior and such cyclic behavior has to follow the same periodicity as the higher level meta-pattern by sufficient number of times (i.e., at least min_rep times).*

In the above example, the meta-pattern X_1 can serve as a component of a higher level meta-pattern (e.g., X_2) only if the locations of valid subsequences of X_1 exhibits a cyclic behavior with a period equal to the span of X_2 (i.e., 150). Otherwise, X_1 could not serve as a component of X_2. This property suggests that we can avoid the generation of a huge number of unnecessary candidate meta-patterns by deriving candidates from qualified span-component combinations according to the component location property. To identify qualified span-component combinations, we need to detect the periodicities exhibited by the locations of valid subsequences of each low level meta-pattern. This can be achieved without generating all valid subsequences for a meta-pattern. In fact, only the set of **maximum valid segments** are sufficient. For a given pattern, a valid segment is a maximum valid segment if it is not a portion of another valid segment. For example, if $min\_rep = 3$ and $max\_dis = 6$, $\{S_1, S_2, S_3, S_4, S_5 S_6\}$ is the set of maximum valid segments of basic pattern $(a, *)$ for the symbol sequence in Figure 3.14(a). Usually, the number of maximum valid segments is much smaller than the number of valid subsequences. The total number of distinct valid subsequences of $(a, *)$ in the symbol sequence given in Figure 3.14(a) would be in the order of hundreds. It is in essence an exponential function of the number of maximum valid segments. Furthermore, for each maximum valid segment, we only need to store a pair of location indexes indicating its starting and ending positions. In the above example, the segment S_1 occupies 8 positions (positions 1 to 8) in Figure 3.14(a) and its location indexes is the pair $(1, 8)$. The location indexes of maximum valid segments indeed provide a compact representation of all necessary knowledge of a valid low level meta-pattern and is motivated by the following observations.

- Given the set of location indexes of maximum valid segments of a pattern, it is easy to compute all possible starting positions and ending positions of valid subsequences. Any starting position of a valid segment is also a starting position of a valid subsequence because a valid subsequence is essentially a list of valid segments. Given a maximum valid segment S containing r repetitions of the pattern, there are $r - min\_rep + 1$ distinct starting positions that can be derived from S. More specifically, they are the positions of the first $r - min\_rep + 1$ occurrences of the pattern in S, respectively. For instance, positions 1 and 3 are the two starting positions derived from S_1. Similarly, all possible ending positions can be computed as well.

- The starting positions of the valid subsequences that exhibit cyclic behavior also present the same periodicity and so do their ending positions. Figure 3.14(b) shows the set of possible starting positions and ending positions of valid subsequences of $(a, *)$. When $min\_rep = 3$, by careful examination, the potential periodicities of $(a, *)$ (i.e., the possible spans of meta-patterns that $(a, *)$ may participate in as a component) include 7, 9, 11, 18, and 20. The periodic behavior discovered on starting positions and ending positions for span = 18 is shown in Figure 3.14(c) and (d), respectively.

Thus, our strategy is to first compute the set of possible starting positions and identify, if any, the ones that exhibit some periodicity. The same procedure is also performed on the ending positions. If the same periodicity exists for both starting positions and ending positions, we then examine, for each pair of starting and ending positions, whether a valid subsequence exists and what

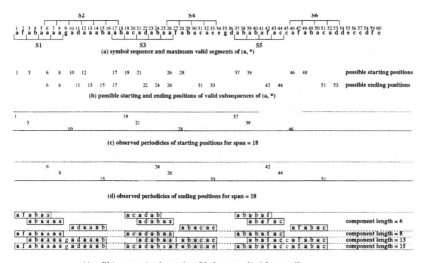

Figure 3.14.   Computation-based Approach

is the possible format of the higher level meta-pattern (i.e., possible span of
the meta-pattern and possible length of the component). Figure 3.14(e) shows
some candidate components generated from $(a, *)$ and the valid subsequences
that support them. It is important to notice that the maintenance of the location
indexes of maximum valid segments leads to a double-win situation. Besides
its positive role in candidate generation, it also enables the verification process
to be accomplished efficiently without the expensive generation and maintenance of all valid subsequences nor the necessity of resort to the raw symbol
sequence. As a result, we devise an efficient **computation-based** algorithm
(as opposite to the traditional match-based approach) in the sense that the discovery of valid meta-patterns (other than base patterns) can be accomplished
through pure computation (performed on the location indexes of maximum
valid segments) without ever resort back to the raw symbol sequence. It has
been demonstrated that this advantage offers at least two orders of magnitudes
speed-up comparing to the match-based approach. The component location
property can provide substantial inter-level pruning effect during the generation of high level candidates from valid low level meta-patterns; whereas the
traditional Apriori property can render some pruning power to conduct the mining process of meta-patterns of the same level. While all meta-patterns can be
categorized according to their levels and the number of non "*" components in
the pattern as shown in Figure 3.15, the pruning effects provided by the component location property and the Apriori property are indicated by dashed arrows

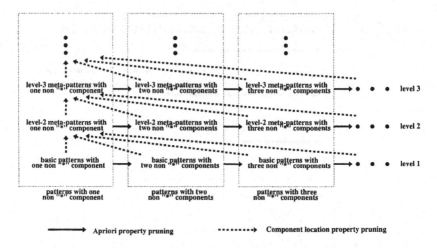

Figure 3.15. Pruning Directions

and solid arrows, respectively. Consequently, the algorithm consists of two levels of iteration. The outer iteration exploits the component location property while the inner iteration utilizes the Apriori property. More specifically, each outer iteration discovers all meta-patterns of a certain level (say, level l) and consists of the following two phases.

1 *candidate component generation.* For each newly discovered valid pattern (or meta-pattern) of level l, generate candidate components for meta-patterns of level $l + 1$. The component location property is employed in this phase.

2 *candidate pattern generation and verification.* This phase generates candidate meta-patterns of level $l + 1$ based on the candidate components discovered in the previous step and validates them. This phase utilizes the Apriori property and contains an iteration loop. During each iteration, meta-patterns with a certain number (say k) of non "*" components are examined, which includes the following two steps.

   (a) If $k = 1$, the candidate singular meta-patterns[9] of level $l + 1$ are generated from candidate components derived in the previous phase. Otherwise, the candidate meta-patterns of level $l + 1$ with k non "*" components are generated based on the discovered level-$(l + 1)$ meta-patterns with $(k - 1)$ non "*" components.

   (b) The newly generated candidate patterns are validated.

This inner iteration continues until no new candidate patterns of level $(l+1)$ can be generated.

The entire procedure terminates when no new candidate components can be generated. In the following sections, we will elaborate on each phase in detail.

Candidate Component Generation

For a given pattern X of level l, a three-step algorithm is devised to generate candidate components (based on X) of patterns of level $l + 1$. The formal description of the algorithm is in $Component\_Generation$.

/* To find component candidates for patterns of level $l + 1$ */

Component_Generation(min_rep, Seg_X)
/* Seg_X is the set of maximum valid segments for an lth level pattern X */
{
    $starting \leftarrow \emptyset$
    /* It holds possible starting positions of valid subsequences of X. */
    $ending \leftarrow \emptyset$
    /* It holds possible ending positions of valid subsequences of X. */
    $start \leftarrow \emptyset$
    /* It holds starting positions in $starting$ that exhibit some periodicity. */
    $end \leftarrow \emptyset$
    /* It holds ending positions in $ending$ that exhibit some periodicity. */
    $candidates \leftarrow \emptyset$
    /* It holds the set of valid subsequences (of X) that support candidate components derived from X. */
    Find_Start_End($Seg_X, starting, ending$)
    Periodicity($starting, start$)
    /* Finding positions in $starting$ that exhibits some span s
    and storing them into $start$ structures. */
    Periodicity($ending, end$)
    /* Finding positions in $ending$ that exhibits some span s
    and storing them into end structures. */
    for each retained span s **do**
        Component ($s, start, end, candidates$)
        /* For a given span s, find the possible length of the component
        that involves X */
    return ($candidates$)
}

The proposed algorithm assumes that for a given lth level pattern X, the set of the maximum valid segments of X are stored in Seg_X. Note that only the location indexes are physically stored and all other information associated with each maximum segment can be derived easily. For example, $S_1, S_2, S_3, S_4, S_5,$

and S_6 are the maximum valid segments of $(a, *)$ in Figure 3.14(a). Looking ahead, in the meta-pattern validation procedure (see Section Canonical Pattern), the algorithm will output all maximum valid segments of a valid meta-pattern as a by-product of the longest valid subsequence discovery without incurring any extra overhead. Note that we only keep track of the set of maximum valid segments instead of all valid subsequences because the number of the valid subsequences is much larger than the number of the maximum valid segments. The benefits of only tracking maximum valid segments are shown in the empirical study section.

In the subroutine $Find\_Start\_End$, each maximum valid segment (in Seg_X) is examined successively to compute the set of possible starting and ending positions of valid subsequences. A possible starting position is the one that has at least min_rep repetitions immediately following it. Similarly, a possible ending position is the one with at least min_rep repetitions immediately proceeding it. Figure 3.14(b) shows the possible starting/ending positions derived from the maximum valid segments in Figure 3.14(a).

Even though two maximum segments in Seg_X may overlap, their $start\_rep$ sets are disjoint from each other (that is, no position (in the input sequence) serves in the set $start\_rep$ for multiple maximum segments). For instance, S_1 and S_2 overlap with each other, and $\{1, 3, 5, 7\}$ and $\{6, 8, 10, 12, 14, 16\}$ are the sets of starting positions of repetitions in S_1 and S_2 respectively. They are obviously disjoint. Therefore, each position in the input sequence serves as a starting position at most once. The same proposition also holds for the ending position. Thus, the computational complexity of each invocation of $Find\_Start\_End$ is $O(N)$ where N is the length of the input sequence.

Find_Start_End(Seg_X, $starting$, $ending$)
{
    for each segment $seg \in Seg_X$ **do**
      for each $start\_rep \in seg$ **do**
      /* $start\_rep$ records the starting position of each repetition
      in a maximum segment. */
        if (Valid_Start($start\_rep$, seg)) **do**
      /* If there are at least min_rep repetitions following $start\_rep$,
      then Valid_Start returns true, otherwise false. */
          $starting \leftarrow starting \cup start\_rep$
      for each $end\_rep \in seg$ **do**
      /* end_rep records the ending position of each repetition
      in a maximum segment. */
        if (Valid_End(end_rep, seg)) **do**
        /* If there are at least min_rep repetitions preceding end_rep,

then Valid_End returns true, otherwise false. */
$$ending \leftarrow ending \cup end\_rep$$
return

}

The $Periodicity$ function locates periodicity presented in a list of positions and is applied to both starting positions and ending positions. With two scans of the list of possible starting positions, we can generate potential span s of higher level meta-pattern that X may participate in. Intuitively, s should be at least the minimum length of a valid subsequence of X (i.e., $span(X) \times min\_rep$ where $span(X)$ is the span of X). For example, in Figure 3.14(a), the minimum length of a valid subsequence of $(a, *)$ is $2 \times 3 = 6$ if $min\_rep = 3$. In turn, the span of any valid meta-pattern that takes $(a, *)$ as a component has to be at least 6 since the component $(a, *)$ will occupy at least 6 positions. During the first scan, a counter is initialized for each possible span greater than or equal to $span(X) \times min\_rep$. While scanning through the list of starting positions, for each starting position x, consider the distance of x to any previous starting position y in the list. If it is larger than $span(X) \times min\_rep$, the corresponding counter is incremented. At the end of the scan, only spans whose corresponding counter reaches $min\_rep - 1$ are retained. During the second pass through the list, for each retained span s, the algorithm locates series of at least min_rep starting positions, such that the distance between any pair of consecutive ones is s. Note that there may be multiple series existing concurrently. As shown in Figure 3.14(c), $\{1, 19, 37\}$, $\{3, 21, 39\}$, and $\{10, 28, 46\}$ are three series of starting positions that exhibit periodicity of span 18. The same process is also performed on the ending positions. Figure 3.14(d) shows the result for span 18 in the above example.

It is easy to see that the first scan through the list takes $O(|list|^2)$ computations and the second scan may consume $O(|S| \times |list|^2)$ time, where $|list|$ and $|S|$ are the number of positions in $list$ and the number of retained spans after the first scan, respectively. Since the number of positions in $list$ is at most the length of the input sequence, the overall computational complexity is $O(|S| \times |list|^2) = O(|S| \times N^2)$ where N is the length of the input sequence.

Periodicity($list, potential$)
{
    for each potential span s $(s \geq min\_rep \times span(X))$ **do**
        initialize a counter $count[s]$ to zero
    for each position $x \in list$ **do**
        for each position $y \in list$ $(y < x)$ **do**
            $distance \leftarrow x - y$
            increment $count[distance]$

$S \leftarrow$ all spans s where $count[s] \geq min\_rep - 1$
/* S holds spans of potential periodicities that may exhibit in $list$ */
for each span $s \in S$ **do**
    for each position $x \in list$ **do**
        if there exists a series of positions $x, x + s, \ldots, x + i \times s$ in $list$
            and $i \geq min\_rep$ **do**
        /* In the case that multiple valid values of i exist, the maximum
        one is chosen. */
            $potential[s] \leftarrow potential[s] \cup \{x, x + s, \ldots, x + i \times s\}$
            /* $potential$ stores the set of positions in $list$ that
            exhibit periodicity. */
    return
}

Then, in the function $Component$, for each retained span s, consider each pair of cyclic starting position series $x, x + s, \ldots, x + s \times j_x$ and cyclic ending position series $y, y + s, \ldots, y + s \times j_y$. If there exist two sub-series: $x', x' + s, \ldots, x' + \lambda \times s$ and $y', y' + s, \ldots, y' + \lambda \times s$ such that $min\_rep \times span(X) \leq y' - x' \leq s$ and $\lambda \geq min\_rep$, then X may participate in a meta-pattern of span s as a component that occupies $y' - x'$ positions. The subsequence from position x' to y', from position $x' + s$ to $y' + s$, and so on are potentially valid subsequences to support X as a component in a higher level meta-pattern. For example, there are 3 series of cyclic starting positions and 3 series of cyclic ending positions corresponding to span 18 as shown in Figure 3.14(c) and (d) respectively. Let's take a look at the starting position series $3, 21, 39$ and ending position series $15, 33, 51$. The starting position 3 and ending position 15 satisfy the above condition and the subsequences from position 21 to 33, and from position 39 to 51 are two additional potential valid subsequences to support $(a, *)$ as a component of a higher level meta-pattern and such component occupies 13 positions (Figure 3.14(e)). Note that the subsequences in this example are not perfect repetitions of $(a, *)$. In fact, each of them consists of two perfect segments of $(a, *)$ separated by a disturbance of length 1. This example further verifies that the meta-pattern can indeed tolerate imperfectness that is not allowed in basic periodic patterns.

Since for a given span s, the cardinalities of $start[s]$ and $end[s]$ are at most the input sequence length N, the computational complexity of $Component()$ is $O(|start[s]| \times |end[s]|) = O(N^2)$ for a given span s.

$Component(s, start, end, candidates)$
{
    for each $start\_position \in start[s]$ with span s **do**
        for each $end\_position \in end[s]$ with span s **do**

if $(end\_position - start\_position \leq s)$ **and**
    /* The total length of the component cannot exceed
    the span of the periodicity. */
    $(end\_position - start\_position \geq min\_rep \times span(X))$ **and**
    /* The length of the component has to be at least the minimum
    length of a valid subsequence of X */
    (Valid_Subsequence $(start\_position, end\_position))$ **do**
    /* Valid_Subsequence returns *true* if the subsequence between
    $start\_position$ and $end\_position$
    is a valid subsequence of X. */
{

$component\_length \leftarrow end\_position - start\_position$
$new\_component \leftarrow (X, component\_length)$
$candidates \leftarrow candidates \cup$
    $(start\_position, end\_position, new\_component)$

}

}

Notice that the above identified potential subsequences are not guaranteed to be valid because we only consider the possible starting and ending positions and ignore whether the symbols in between form a valid subsequence. In fact, the identified subsequences might not be valid especially in the scenario where valid segments scatter sparsely throughout the data sequence. This can be observed from the example in Figure 3.16. The three potential subsequences generated for component length 18 are invalid if $max\_dis = 5$. Therefore, it is necessary to validate these subsequences. We note that the set of maximum valid segments and their connectivity[10] can be organized into a graph and a depth first traversal would be able to verify whether a subsequence between two positions is valid or not.

It is easy to see that each invocation of $Component\_generation()$ would take $O(|S| \times N^2)$ computation. Note that this is only a worst case bound and the actual running time is usually much faster. We also want to mention that, without the component location property, an algorithm that employs the Apriori property would have to consume as much as $O(N^{2 \times min\_rep})$ time. Experiment studies in the empirical study section also demonstrates the pruning power of component location property.

Candidate Pattern Generation and Validation

Candidate Pattern Generation

For each combination of component P, span s, and component length p, the candidate singular pattern $(P[1, p], * : [p + 1, s])$ is constructed. In the previous example in Figure 3.14, four candidate singular meta-patterns (starting

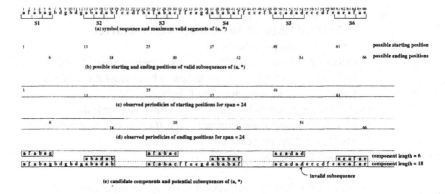

Figure 3.16. An example of invalid potential subsequence

at position 1) are constructed from the candidate component $(a, *)$ for span 18, one for each distinct component length shown in Figure 3.14(e). They are $((a, *) : [1, 6], * : [7, 18]), ((a, *) : [1, 8], * : [9, 18]), ((a, *) : [1, 13], * : [14, 18])$, and $((a, *) : [1, 15], * : [16, 18])$. Note that any pattern of format $(* : [1, t], P[t+1, t+p], * : [t+p+1, s])$ is essentially equivalent to $(P[1, p], * : [p + 1, s])$ with a shifted starting position in the data sequence.

For the generation of candidate complex patterns, the Apriori property is employed. The candidate pattern $(x_1 : [1, t_1], x_2 : [t_1 + 1, t_2], \ldots, x_l : [t_{k-1} + 1, s])$ is constructed if all of $(x_2 : [1, t_2 - t_1], \ldots, x_l : [t_{k-1} - t_1 + 1, s - t_1], * : [s - t_1 + 1, s])$[11], $(x_1 : [1, t_1], * : [t_1 + 1, t_2], \ldots, x_l : [t_{k-1} + 1, s]), \ldots,$ and $(x_1 : [1, t_1], x_2 : [t_1 + 1, t_2], \ldots, * : [t_{k-1} + 1, s])$ are valid patterns. Referring back to the inventory replenishment example discussed previously. After we identify $((r, *) : [1, 24], * : [25, 52])$ and $((r, * : [2, 3]) : [1, 27], * : [28, 52])$ as valid patterns (through the process presented previously), two candidate patterns will be constructed via the Apriori property, and they are $((r, *) : [1, 24], *, (r, * : [2, 3]) : [26, 52])$ and $((r, *) : [1, 24], (r, * : [2, 3]) : [25, 51], *)$. In general, for a given set of valid patterns, multiple candidate patterns can be constructed, each of which corresponds to a possible layout of gaps (filled by *) between each pair of consecutive non "*" components. This is the primary difference between the application of Apriori property in traditional models and in the discovery of meta-pattern.

Canonical Pattern

The candidate patterns generated by the method above may be redundant. For instance, let's consider sequence ABABABAB. Both patterns (AB) and (ABAB) can be valid patterns if $min\_rep \leq 3$. However, it is clear that the pattern is (ABAB) is a redundant pattern of (AB). Thus, all valid patterns can be divided into two categories: derived patterns and canonical patterns.

Definition A pattern P_1 is called a **derived pattern** if there exists another pattern P_2 ($P_2 \neq P_1$) and P_1 can be formed by append multiple P_2 together.

We say P_1 is derived from P_2. In the previous example, (ABAB) is derived from (AB).

Definition A pattern P_1 is called a canonical pattern if there does not exist any pattern from which P_1 is derived.

(AB) is a canonical pattern in the previous example.

In many applications that users are only interested in the valid canonical patterns. To remove the derived patterns from consideration, we use the following observation. Let P_1 is a pattern derived from P_2. P_1 is valid if P_2 is valid by definition. As a result, we only need to consider the canonical patterns. During candidate generation phase, we take a look a candidate to see whether it is canonical. If so, we keep it as a candidate. Otherwise, it is removed from further consideration.

Candidate Pattern Validation

For a given candidate meta-pattern X, we need to find the longest valid subsequence for X if there exists any. This is similar to the longest subsequence identification algorithm for asynchronous patterns [9] with one exception: a component in a meta-pattern may correspond to a valid subsequence of some lower level pattern, while each component for an asynchronous pattern is restricted to a symbol. This brings some difficulty to the identification of occurrence of X in the symbol sequence. Fortunately, the problem can be easily solved since we keep in *candidate* the set of maximum valid segments that may support a lower level pattern as a component of X in the step of candidate component generation. To verify whether a component Y matches a certain portion of the sequence, we can go back to verify Y against the maximum valid segments stored in *candidate*. To further improve the efficiency of the algorithm, the segments in *candidate* can be organized in some data structure, e.g., a tree.

The rest of algorithm deals with how to stitch occurrences of meta-pattern together to generate the longest valid subsequence. This part of the algorithm is exactly the same as that for mining asynchronous patterns where, for a given candidate meta-pattern X, the longest valid subsequence can be located in linear computation time with respect to the length of the symbol sequence. Note that this procedure does not need to resort to the raw symbol sequence and therefore can be accomplished very efficiently.

At the same time of generating the longest valid subsequence, a separate data structure Seg_X is maintained simultaneously to store all maximum valid segments. Every time a match of the meta-pattern X is discovered, if there exists a segment seg in Seg_X such that the last repetition of X in seg is adjacent to the current match, then we extend seg to include the current match. Otherwise, a new segment that consists of the current match is created. After a scan of the input symbol sequence, segments that contain less than min_rep repetitions of X are removed from Seg_X. The remaining segments are the maximum valid segments of X, which will be used in generating candidate components of higher level meta-patterns.

Each invocation of $Maximum\_Valid\_Segment()$ takes $O(N)$ time to finish. The set Seg_X can be indexed according to the ending position of each segment to facilitate the process. Since there are at most $\frac{N}{span(X) \times min\_rep}$ segments of X, the space required to store Seg_X is $O(\frac{N}{span(X) \times min\_rep})$. Note that we only need to store the starting and ending positions of each segment.

Maximum_Valid_Segment()
{
    for each match M of X discovered in the symbol sequence **do**
        if there exist a segment $seg \in Seg_X$ s.t. seg is adjacent to M **do**
            extend seg to include M
        else
            $newseg \leftarrow M$
            $Seg_X \leftarrow Seg_X \cup \{newseg\}$
    for each $seg \in Seg_X$ **do**
        if seg has less than min_rep repetitions **do**
            $Seg_X \leftarrow Seg_X - \{seg\}$
    return
}

Experimental Results for Meta-Pattern

The meta-pattern discovery algorithm is implemented in C on an AIX workstation with 300 MHz CPU and 128 MB main memory. A real trace log from the search engine *scour.net* is employed to evaluate the benefits of the meta-pattern model while four synthetically generated sequences are used to measure the performance of our algorithm.

min_rep	3	10	20
max_dis	200	200	200
Level-1 Patterns	107	31	15
Level-2 Patterns	12	5	2
Level-3 Patterns	1	0	0
Meta only Patterns	10	5	2

Table 3.2. Patterns Discovered in Scour Trace

Scour Traces

Scour is a web search engine specialized in multimedia content search whose URL is "http://www.scour.net". Since early of 2000, the average daily number of hits on Scour has grown over one million. A trace of all hits on Scour between March 1 and June 8 (total 100 days) [1] were collected. The total number of accesses is over 140 million. Then the entire trace is summarized into a sequence as follows. The trace is divided into 30 minute intervals. The number of hits during each 30 minute interval is calculated. Finally, we label each interval with a symbol. For example, if the number of hits is between 0 and 9999, then this interval is labeled as a, if the number of hits is between 10000 and 19999, then this interval is labeled as b, and so on. The summarized sequence consists of 4800 occurrences of 43 distinct symbols.

Table 3.2 shows the number of patterns discovered from the Scour sequence with respective thresholds. There exist some interesting patterns. When min_rep and max_dis are set to 3 and 200, respectively, there is a level 3 meta-pattern. This level 3 pattern describes the following phenomenon. In a weekday between 4am and 12pm EST, there exists a pattern (b,b) where b stands for the number of hits is between 10000 and 19999; and during 5pm to 1:30am EST, we found the pattern $(e, *)$, which means that the number of hits is between 40000 and 49999. Furthermore, this pattern repeated itself during each weekday within a week (i.e., level-2 pattern) and it also exhibits weekly trend (i.e., level-3 pattern). This observation confirms with the cyclical behavior of the Internet traffics discovered in [7]. Furthermore, various studies[2] have shown that the traffic on the world wide web exhibits self-similarity which also confirms our findings. In addition, we also compute the meta-patterns that can not be expressed in the form of basic patterns. We call these patterns *meta only* patterns and the number of these patterns is also shown in Table 3.2. From this table, we can see that most of the discovered level 2 and 3 patterns can not be expressed in the form of basic patterns, and thus can only be represented as meta-patterns.

To further understand the behavior of our proposed meta-pattern mining algorithm, we constructed four long synthetic sequences and the performance of our algorithm on these sequences is presented in the following section.

Data Set	μ_l	μ_s	μ_r	μ_d
$DS1$	5	5	50	50
$DS2$	5	5	1000	1000
$DS3$	100	1000	50	50
$DS4$	100	1000	1000	1000

Table 3.3. Parameters of Synthetic Data Sets

min_rep	Scour trace	$DS1$	$DS2$	$DS3$	$DS4$
10	0.008	0.002	0.002	0.001	0.003
20	0.002	0.0008	0.0007	0.0002	0.0009
30	0.0008	0.0002	0.0003	0.00008	0.0002
40	0.0003	0.00004	0.00007	0.00001	0.00005

Table 3.4. Effects of min_rep on pruning power

Synthetic Sequences

The four synthetic sequences are generated as follows. Each sequence consists of 1024 distinct symbols and 20M occurrences of symbols. The synthetic sequence is generated as follows. First, at the beginning of the sequence, the level of pattern is determined randomly. There are four possible levels, i.e., 1, 2, and 3, 4. Next, the number of segments in this pattern is determined. The length l of each segment is selected based on a geometric distribution with mean μ_l. The number of repetitions of a lower level pattern in a segment is randomly chosen between min_rep and $\lfloor \frac{l}{p} \rfloor$ where p is the span of the lower level pattern. The number of symbols involved in a pattern is randomly chosen between 1 and the span p. The number of valid segments is chosen according to a geometrical distribution with mean μ_s. After each valid segment, the length of the disturbance is determined based on a geometrical distribution with mean μ_d. This process repeats until the length of the sequence reaches 20M. Four sequences are generated based on values of μ_l, μ_s, μ_r, and μ_d in Table 3.3.

Effects of Component Location Property Pruning

In our proposed algorithm, we use the component location property pruning to reduce the candidate patterns. Table 3.4 shows the pruning power of our algorithm. The pruning power is measured as the fraction of candidate patterns. We can see that the candidate patterns in our algorithm is around 10^{-2} to 10^{-4} of the overall patterns. This means that on average less than 1% of all patterns need to be examined in our algorithm. In addition, the pruning power increases (i.e., the fraction decreases) with larger min_rep because less patterns may be qualified by a larger (more restricted) min_rep parameter. In this experiment, the max_dis is fixed to 20.

max_dis	Scour trace	$DS1$	$DS2$	$DS3$	$DS4$
10	0.0009	0.0002	0.0002	0.0001	0.0003
20	0.002	0.0008	0.0007	0.0002	0.0009
30	0.008	0.002	0.003	0.0008	0.002
40	0.01	0.007	0.009	0.002	0.005

Table 3.5. Effects of max_rep on pruning power

Level	$DS1$	$DS2$	$DS3$	$DS4$
2	0.02	0.02	0.04	0.01
3	0.05	0.03	0.01	0.02

Table 3.6. CPU Time Ratio of Computation-Based Approach vs. Match-Based Approach

We also study the effects of the parameter max_dis on our algorithm. Table 3.5 shows the effects of max_dis. The pruning power of our algorithm is evident. More than 99% of patterns are pruned. The pruning power decreases with larger max_dis threshold because more patterns may be qualified. Overall, the component location property can prune away a significant amount of patterns and thus reduce the execution time of the pattern discovery process. We fix min_rep to 20.

Effects of Computation-Based Approach

In our approach, we only store the maximum valid segments for each pattern. We compare the computation-based approach with the match-based approach. In the match-based approach, for each pattern, all valid subsequences are stored and used to mine higher level meta-patterns. For each level of pattern, we track the CPU time consumed by the computation-based approach and the match-based approach. (We assume that all information can be stored in main memory. Since the number of possible subsequences is much larger than that of maximum valid segments, the match-based approach has more advantages with this assumption.) The ratio of the CPU time of the computation-based approach over that of the match-based approach is calculated and presented in Table 3.6. It is obvious that the computation-based approach can save at least 95% of the CPU time comparing to the match-based approach. This is due to the fact that the number of maximum valid segments is far less than that of valid subsequences as we explained in Section 5.

Overall Response Time

The overall response time is one of the most important criterion for evaluation of an algorithm. We mine the meta-patterns with different min_rep threshold. For a given min_rep, we mine the patterns on all four data sets and the average

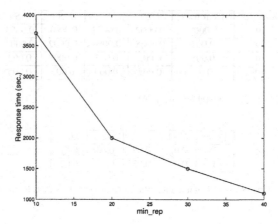

Figure 3.17.   Response Time of Meta-pattern Mining Algorithm

response time over the four data sets are taken and shown in Figure 3.17. The average response time decreases exponentially as min_rep increases. Although the average response time is a bit long when min_rep is small, but it is still tolerable (around 1 hour) due to the pruning effects of component location property and the computation-based candidate pattern generation. The meta-pattern mining algorithm is also applied to symbol sequences with different lengths. We found that the response time of our algorithm is linearly proportional to the length of the symbol sequence.

Notes

1 The replenishment order of a merchandize may not be prescheduled but rather be filed whenever the inventory is low.
2 It is obvious that there can be at most l different ending positions for these subsequences.
3 In other words, seq is longest subsequence among those with ending position prior to i.
4 No new subsequence needs to be created if seq ends at position $i-1$ because all necessary extensions of subsequences ending at position $(i-1)$ have been taken at Step 2(b). Step 3 is essentially designed to give the valid subsequence(s) that ends prior to position $(i-1)$ the opportunity to be extended further by appending the current match.
5 Otherwise, all subsequences in that entry would be extended to end at position $k+l$.
6 Note that in many applications, e.g., seismic periodicity analysis, the period length is usually unknown in advance and is part of the mining objective.

7   It is introduced to represent the position(s) in a pattern where no strong periodicity exhibits.

8   Since each component corresponds to exactly one symbol, we do not have to explicitly record the offset of a component within the pattern as this information can be easily derived.

9   A singular meta-pattern is a meta-pattern that has only one non "*" component. Otherwise, it is called a complex meta-pattern.

10 If the disturbance between two segments is at most max_dis, then we consider they are connected. Otherwise, they are considered not connected.

11 This is equivalent to $(* : [1, t_1], x_2 : [t_1 + 1, t_2], \ldots, x_l : [t_{k-1} + 1, s])$.

References

[1] Busam, V. (2000). Personal Communications.

[2] Crovella, M., and Bestavros, A. (1997) Self-similarity in world wide web traffic: evident and possible cause. *IEEE/ACM Transactions on Networking.* 5(6):835-846.

[3] Han, J., Gong, W., and Yin, Y. (1998) Mining segment-wise periodic patterns in time-related databases. *Proc. Int'l. Conference on Knowledge Discovery and Data Mining (KDD).* pp. 214-218.

[4] Han, J., Dong, G., and Yin, Y. (1999). Efficient mining partial periodic patterns in time series database. *Proc. IEEE Int'l. Conference on Data Engineering (ICDE).* pp. 106-115.

[5] Ozden, B., Ramaswamy, S., and Silberschatz, A. (1998). Cyclic association rules. *Proc. 14th Int'l. Conference on Data Engineering (ICDE).* pp. 412-421.

[6] Ramaswamy, S., Mahajan, S., and Silberschatz, A. (1998). On the discovery of interesting patterns in association rules. *Proc. 24th Intl. Conf. on Very Large Data Bases (VLDB).* pp. 368-379.

[7] Thompson, K., Miller, G., and Wilder, R. (1997). Wide-area internet traffic patterns and characteristics. *IEEE Network - Magazine of Global Information Exchange,* 11(6):10-23.

[8] Wang, W., Yang, J., and Yu, P. (2001) Meta-patterns: revealing hidden periodic patterns. *IEEE Int'l. Conference on Data Mining (ICDM).* pp. 550-557.

[9] Yang, J., Wang, W., and Yu, P. (2003). Mining asynchronous periodic patterns in time series data. *IEEE Transactions on Knowledge and Data Engineering.* 15(3):613-628.

Chapter 4

STATISTICALLY SIGNIFICANT PATTERNS

In many applications, e.g., bioinformatics, system traces, web logs, etc., the occurrences of symbols in a sequence may follow a very skewed distribution. For instance, in a DNA sequence, the occurrences of GC is much higher than those of AT within coding regions. By chance, a pattern with common symbols would occur much more frequently than that with rare symbols. Therefore, the frequency is not a good measurement of the importance of a pattern. To overcome this problem, we will discuss a new model to capture the importance of a sequential pattern, *statistically significant Sequential Patterns*. We will first describe an algorithm for mining statistically significant patterns within the entire sequence [18] and later we will present an algorithm for mining the patterns that are statistically significant in some subsequence [19].

1. InfoMiner

A periodic pattern is an ordered list of events which repeats itself in the event sequence. It is useful in characterizing the cyclic behavior. As a newly developed research area, most previous work on mining sequential data addresses the issue by utilizing the support (number of occurrences) as the metric to identify important patterns from the rest [11, 12, 16, 20]. A qualified pattern in the support model must occur sufficient number of times. In some applications, e.g., market basket, such a model has proved to be meaningful and important. However, in other applications, the number of occurrences may not always represent the significance of a pattern. Consider the following examples.

- *Computational Biology.* A genome consists of a long sequence of nucleotides. Researchers are interested in motifs (i.e., (short) subsequences of nucleotides) which are statistically significant rather than those occurring frequently. The statistical significance of a motif is defined as how likely such a nucleotide permutation would occur in an equivalent random data sequence [7]. In other words, we want to find the motifs that occur at a frequency higher than their expected frequencies. It is obvious that a statistically significant motif may not necessarily occur frequently, thus, the support metric is not an appropriate measurement of the significance of a motif.

■ *Web server load.* Consider a web server cluster consisting of 5 servers. The workload on each server is measured by 4 ranks: *low*, *relatively low*, *relatively high*, and *high*. Then there are $4^5 = 1024$ different events, one for each possible combination of server states. Some preliminary examination of the cluster states over time might show that the state fluctuation complies with some periodic behavior. Obtaining such knowledge would be very beneficial for understanding the cluster's behavior and improving its performance. Although having high workload on all servers may occur at a much lower frequency than other states, patterns involving it may be of more interest to administrators if the occurrence of such patterns contradicts the prior expectation.

■ *Earthquake.* Earthquakes occur very often in California. It can be classified by its magnitude and type. Scientists may be interested in knowing whether there exists any inherent seismic period so that prediction can be made. Note that any unidentified seismic periodicity involving major earthquake is much more valuable even though it occurs at a much lower frequency than minor ones.

In above examples, we can see that users may be interested in not only the frequently occurred patterns, but also the surprising patterns (i.e., beyond prior expectation) as well. A large number of occurrences of an "expected" frequent pattern sometimes may not be as interesting as a few occurrences of an "expected" rare pattern. The support model is not ideal for these applications because, in the support model, the occurrence of a pattern carries the same weight (i.e., 1) towards its significance, regardless of its likelihood of occurrence. Intuitively, the assessment of significance of a pattern in a sequence should take into account the expectation of pattern occurrence (according to some prior knowledge). Recently, many research has been proposed [3, 6, 8, 9, 13, 14, 15, 17, 21] towards this objective. In this section, a new model is proposed to characterize the class of so-called *surprising* patterns (instead of frequent patterns). We will show that our model not only has solid theoretical foundation but also allows an efficient mining algorithm.

The measure of *surprise* should have the following properties. (1) The surprise of a pattern occurrence is anti-monotonic with respect to the likelihood that the pattern may occur by chance (or by prior knowledge). (2) The metric should have some physical meaning, i.e., not arbitrary created. It is fortunate that the *information* metric [5] which is widely studied and used in the communication field can fulfill both requirements. Intuitively, information is a measurement of how likely a pattern will occur or the amount of "surprise" when a pattern actually occurs. If a pattern is expected to occur frequently based on some prior knowledge or by chance, then an occurrence of that pattern carries less information. Thus, we use information to measure the surprise of an occurrence of a pattern. The *information gain* metric is introduced to represent the accumulated information of a pattern in an event sequence and is used to assess the degree of surprise of the pattern. In the remainder of this chapter, we refer to this model as the *information model*.

The information model is different from the support model. For a given minimum information gain threshold, let Ψ be the set of patterns that have information gain higher than this threshold. Under the support model, in order to find all patterns in Ψ when event occurrence frequencies are vastly different, the minimum support threshold has to be set very low. A major problem could rise from this: too many patterns. Table 4.1 shows a comparison between the support model and the information model. The test sequences are constructed from real traces. (The construction of the sequence is described in the experimental section.) In order to find the pattern with the highest information gain, the support threshold has to be set at 0.000234 and there are over 16,000 satisfied patterns in one sequence. It is obvious that the support threshold has to be set very low to discover a small number of patterns with high information gain. This means that patterns with very high information gain are buried in a sea of patterns with relatively low information gain. This could be a large burden for the end user to distinguish the surprising patterns (i.e., patterns with high information gain) from the rest. In addition, since a large number of patterns have to be generated, the support model may yield an inefficient algorithm.

Table 4.1. Support threshold vs. information gain threshold

Number of Patterns	Scour Trace	
Satisfied Info. Thresh.	Support Thresh.	Num. of satisfied patterns
1	0.000234	16,123
10	0.000212	16,953
100	0.000198	17,876
Number of Patterns	IBM Trace	
Satisfied Info. Thresh.	Support Thresh.	Num. of satisfied patterns
1	0.0035	637
10	0.0031	711
100	0.0024	987

Although the information gain is a more meaningful metric for the problems addressed previously, it does not preserve the *downward closure* property (as the *support* does). For example, the pattern (a_1, a_2) may have enough information gain while neither $(a_1, *)$ nor $(*, a_2)$ does. The $*$ symbol represents the "don't care" position, which is proposed in [11]. We cannot take advantage of the standard pruning technique (e.g., Apriori algorithm) developed for mining association rules [1] and temporal patterns [2, 12]. Fortunately, we are able to identify the *bounded information gain property* where patterns with inextensible prefixes could not be surprising (given some information gain threshold) and can be excluded from consideration at a very early stage. This motivates us to devise a recursive algorithm as the core of our pattern discovery tool, InfoMiner.

Model of Surprising Patterns

We adopt the general model for periodic patterns proposed in [12] with one exception: Instead of finding *frequent* patterns[1], our goal is to discover **surprising** patterns in an event sequence. Let $E = \{a_1, a_2, \dots\}$ be a set of distinct events. The event sequence is a sequence of events in E. A periodic pattern is a list of l events that may occur recurrently in the sequence with period length l. The **information** carried by an event a_i ($a_i \in E$) is defined to be $I(a_i) = -\log_{|E|} Prob(a_i)$ where $|E|$ and $Prob(a_i)$ are the number of events in E and the probability that a_i occurs, respectively. The probability $Prob(a_i)$ can be assessed in many ways which include but are not limited to:

- *Uniform distribution*: $Prob(a_1) = Prob(a_2) = \dots = Prob(a_i) = \dots = \frac{1}{|E|}$;

- *Experimental evidence*: $Prob(a_i) = \frac{Num_D(a_i)}{N}$ for all $a_i \in E$ where $Num_D(a_i)$ and N are the number of occurrences of the event a_i in an event sequence D and the length of D, respectively;

- *Prior belief*: $Prob(a_i)$ is determined by some domain expert.

Without loss of generality, we adopt the second option to assess the information carried by an event, i.e. an occurrence of a frequent event carries less information/surprise than that of a rare event. Note that this also coincides with the original intention of *information* in the data communication community. We shall show later that this gives us the opportunity to handle patterns with divergent probabilities seamlessly. Theoretically speaking, the base of the logarithm function can be any real number that is greater than 1. Typically, $|E|$ is chosen to be the base to play a normalization role in the computation (i.e., $I(a_i) = 1$ if $Prob(a_i) = \frac{1}{|E|}$). For example, the sequence in Figure 4.1 contains 6 different events a_1, a_2, a_3, a_4, a_5, and a_6. Their probabilities of occurrence and information are shown in Table 4.2.

A **pattern** of **length** l is a tuple of l events, each of which is either an event in E, or the **eternal event** (represented by symbol $*$). An eternal event is a

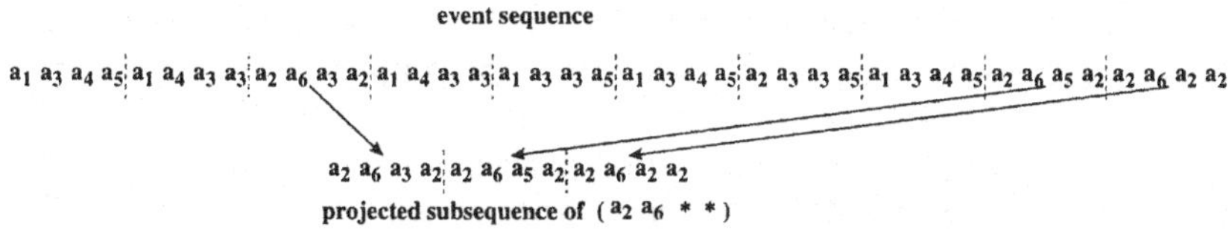

event sequence

a₁ a₃ a₄ a₅ a₁ a₄ a₃ a₃ a₂ a₆ a₃ a₂ a₁ a₄ a₃ a₃ a₁ a₃ a₃ a₅ a₁ a₃ a₄ a₅ a₂ a₃ a₃ a₅ a₁ a₃ a₄ a₅ a₂ a₆ a₅ a₂ a₂ a₆ a₂ a₂

a₂ a₆ a₃ a₂ a₂ a₆ a₅ a₂ a₂ a₆ a₂ a₂
projected subsequence of (a₂ a₆ * *)

Figure 4.1. Event Sequence and Projected Subsequence

Table 4.2. Probability of Occurrence and Information

Event	Probability	Information
a_1	$\frac{6}{40} = 0.15$	$-\log_6(0.15) = 1.06$
a_2	$\frac{8}{40} = 0.20$	$-\log_6(0.20) = 0.90$
a_3	$\frac{12}{40} = 0.30$	$-\log_6(0.30) = 0.67$
a_4	$\frac{5}{40} = 0.125$	$-\log_6(0.125) = 1.16$
a_5	$\frac{6}{40} = 0.15$	$-\log_6(0.15) = 1.06$
a_6	$\frac{3}{40} = 0.075$	$-\log_6(0.075) = 1.45$

virtual event that matches any event in E and is used to represent the "don't care" position in a pattern. By definition, the information of the eternal event $*$ is $I(*) = -log_{|E|} Prob(*) = 0$ since $Prob(*) = 1$. An intuitive interpretation is that the occurrence of an event that is known to be always true does not provide any "new information" or "surprise". A pattern P with length l is in the form of $(p_1, p_2, \ldots, p_l)$ where $p_i \in E \cup \{*\}$ $(1 \leq i \leq l)$ and at least one position has to be filled by an event in E^2. P is called a **singular pattern** if only one position in P is filled by an event in E and the rest positions are filled by $*$. Otherwise, P is referred to as a **complex pattern**. For example, $(*, a_3, *)$ is a singular pattern of length 3 and $(a_2, a_6, *, a_2)$ is a complex pattern of length 4. Note that an event may have multiple occurrences in a pattern. As a permutation of a list of events, a pattern $P = (p_1, p_2, \ldots, p_l)$ will occur with a probability $Prob(P) = Prob(p_1) \times Prob(p_2) \times \cdots \times Prob(p_l)$ in a random event sequence if no advanced knowledge on correlation among these events is assumed. Then the information carried by P is $I(P) = -\log_{|E|} Prob(P) = I(p_1) + I(p_2) + \cdots + I(p_l)$. It follows directly that the information of a singular pattern always equals to the information of the event specified in the pattern. This property provides a natural bridge between events and patterns. For example, $I((*, a_6, *, *)) = I(a_6) = 1.45$ and $I((a_2, a_6, *, *)) = I(a_2) + I(a_6) = 0.90 + 1.45 = 2.35$ according to Table 4.2.

Given a pattern $P = (p_1, p_2, \ldots, p_l)$ and a segment S of l events $s_1, s_2, \ldots, s_l$, we say that S **supports** P if, for each event p_i $(1 \leq i \leq l)$ specified in P, either $p_i = *$ or $p_i = s_i$ is true. The segment a_2, a_6, a_3, a_2 supports the pattern $(a_2, a_6, *, *)$ while the segment a_1, a_6, a_4, a_5 does not. To assess whether a pattern of length l is surprising in an event sequence D, D is viewed as a list of disjoint contiguous segments, each of which consists of l events. The number of segments that support P is also called the **support** of P (denoted by $Support(P)$). The event subsequence[3] consisting of the list of segments that support P is called the **projected subsequence** of D on P. In Figure 4.1, the event sequence consists of 40 events. When mining periodic patterns with $l = 4$, it can be viewed as a list of 10 segments, each of which contains 4 events. The support of $(a_2, a_6, *, *)$ is 3 in the sequence and the projected subsequence on $(a_2, a_6, *, *)$ is $a_2, a_6, a_3, a_2 \, a_2, a_6, a_5, a_2, a_2, a_6, a_2, a_2$. As a measurement of the degree of surprise of a pattern P in an event sequence D, the **information gain** of P in D is defined as $G(P) = I(P) \times (Support(P) - 1)$. Since our objective is to mine surprising periodic patterns, an event combination appearing in the event sequence which never recurs is of little value in this problem domain. Therefore, in the proposed model, only recurrences of a pattern will have positive contribution to the information gain. $Support(P) - 1$ is indeed the number of recurrences of P. In the rest of the chapter, we will use $Repetition(P)$ to denote it. For example, $Repetition((a_2, a_6, *, *)) = 3 - 1 = 2$ and $G((a_2, a_6, *, *)) = 2.35 \times 2 = 4.70$ in Figure 4.1.

Similar to the support model, an information gain threshold, min_gain, is specified by the user to define the minimum information gain to qualify a surprising pattern. Given an event sequence and an information gain threshold, the goal is to discover all patterns whose information gains in the event sequence exceed the threshold. Obviously, the proper value of this threshold is application dependent and may be specified by a domain expert. A user may use the following heuristic to choose the value of min_gain. If a pattern with probability p is regarded as a surprising pattern when it repeats itself by at least n times in the sequence. Then the min_gain can be set to $(-\log_{|E|} p) \times n$ where $-\log_{|E|} p$ is the information of the pattern. Alternatively, the user also has the opportunity to specify the number of (most) surprising patterns needed. We will show later that our proposed algorithm can efficiently produce desired results under both specifications.

It is conceivable that the information model can also be applied to define both surprising itemsets from transaction database and surprising sequential patterns from sequence database. Without loss of generality, we focus our discussion to the domain of mining periodic patterns in a sequence in this chapter. To facilitate the explanation in the rest of the section, we refer to a pattern, say P, as a **subpattern** of another pattern, say P', if P can be generated by replacing some event(s) in P' by the eternal event $*$. P' is called a **superpattern** of P. For example, $(a_2, a_6, *, *)$ and $(*, a_6, *, *)$ are subpatterns of $(a_2, a_6, *, a_2)$. The pattern-subpattern relationship essentially defines a partial order among patterns of the same length.

Projection-based Algorithm

Previous work on pattern discovery usually utilizes the Apriori property that can be stated as "if a pattern P is significant, then any subpattern of P is also significant". This property holds for the support model but is not true in the information model. For example, in Figure 4.1, the information gain $G((a_2, *, *, *)) = 0.90 \times 3 = 2.70$ which is less than the information gain of pattern $(a_2, a_6, *, *)$ (i.e., 4.70). If the threshold is set to be 4.5, then $(a_2, a_6, *, *)$ qualifies while $(a_2, *, *, *)$ does not. (Note that a_6 is an infrequent event which occurs only three times in the event sequence.) This implies that the algorithms developed for the support model are not applicable. The pruning power of the support model essentially comes from the fact that if we know a pattern is not valid, we do not need to examine its superpatterns. Can we achieve a similar pruning power under the information model? To answer this question, we first introduce a concept called *extensible prefix*.

Definition For a pattern $P = (p_1, p_2, \ldots, p_l)$, the tuple $(p_1, p_2, \ldots, p_i)$ is called a **prefix** of P where $1 \le i \le l$.

A prefix is part of a pattern. A pattern can be generated by appending more events to the prefix. For instance, $(a_1, *, a_4)$ is a prefix of patterns $(a_1, *, a_4, a_3)$ $(a_1, *, a_4, a_2)$, $(a_1, *, a_4, a_4)$, and $(a_1, *, a_4, *)$, etc.

Definition Given an information gain threshold min_gain, a prefix is **extensible** if at least one pattern with this prefix is surprising (i.e., whose information gain meets min_gain), and is **inextensible** otherwise.

It follows from the definition that, all prefixes of a surprising pattern are extensible, and any pattern with an inextensible prefix cannot be a surprising pattern. In order to find all surprising patterns, we only need to examine extensible prefixes. The challenge lies on how to recognize inextensible prefixes as early as possible so that they can be excluded from further investigation. However, if all patterns with a certain prefix have to be examined exhaustively before we are able to determine whether the prefix is inextensible, then we will not be able to save any computation. Fortunately, the assessment of extensibility can be done efficiently due to a so-called *bounded information gain* property discussed in the following context.

LEMMA 4.1 *Let $P = (p_1, p_2, \ldots, p_l)$ be a pattern and $P_i = (p_1, p_2, \ldots, p_i)$ be a prefix of P. Given an event sequence, $G(P) \le I(P) \times Repetition(P_i)$, where $I(P) = \sum_{k=1}^{l} I(p_k)$ is the information of the pattern P.*

Proof. Since P_i is a prefix of P, $Repetition(P) \leq Repetition(P_i)$ due to the Apriori property. Thus $G(P) = I(P) \times Repetition(P) \leq I(P) \times Repetition(P_i)$. $\square$

Consider the pattern $P = (a_1, a_3, a_4, a_5)$ and its prefix $P_2 = (a_1, a_3)$ in the sequence in Figure 4.1. Clearly, we have $G(P) = I(P) \times Repetition(P) = 4.34 \times 2 = 8.68$ which is less than the value of $I(P) \times Repetition(P_2) = 4.34 \times 3 = 13.02$. This suggests that we may take advantage of this "bounding" effect when we assess the extensibility of a prefix. For a given prefix P_i, consider the set (denoted by $\Lambda(P_i)$) of patterns of period length l with the same prefix P_i and let P_{max} be the pattern with the highest information in $\Lambda(P_i)$. Then, for any pattern P with prefix P_i (i.e., $P \in \Lambda(P_i)$), the inequality $G(P) \leq I(P_{max}) \times Repetition(P_i)$ holds by Lemma 4.1. Therefore, we can determine whether P_i is extensible by estimating the maximum information $I(P_{max})$ that may be carried by any pattern with prefix P_i. The value of $I(P_{max})$ can be computed by, first, for each unspecified position (following the prefix P_i), identifying the highest information possessed by any potential event for that position; and then aggregating them together with the information possessed by P_i. It is conceivable that the prefix P_i can not be extensible if $I(P_{max}) \times Repetition(P_i)$ is below the information gain threshold min_gain. This leads to the following theorem.

THEOREM 4.2 **(Bounded information gain)** *Given an information gain threshold min_gain and a period length l, a prefix $P_i = (p_1, p_2, \ldots, p_i)$ is not extensible iff $Repetition(P_i) < \frac{min\_gain}{max\_info}$ where $max\_info = I(P_i) + \sum_{k=i+1}^{l} f_k$ is the maximum information that can be carried by any pattern with prefix P_i and f_k is the highest information that can be carried by any potential event at the (unspecified) position k.*

Once a prefix is deemed to be inextensible, we will immediately eliminate it from any further examination. Only extensible prefixes will be used to extend to longer (extensible) prefixes and to construct candidate patterns. Furthermore, given a period length l, for any prefix $P_i = (p_1, p_2, \ldots, p_i)$, consider an unspecified position k where $i < k \leq l$. Not every event can potentially be at position k in a surprising pattern with prefix P_i. An event $e \in E$ can possibly be a candidate for position k only if e recurs on position k for a sufficient number of times. In particular, the minimum required number of repetitions is $min\_rep = \frac{min\_gain}{I(P_i) + \sum_{j=i+1}^{l} f_j}$. This indicates that only a (small) subset of events may serve as the candidates for each unspecified position and we can limit our search to these candidate events, and also leads to the following remark.

REMARK 4.3 **(Candidate refinement)** *For any two prefixes $P_{i1} = (p_1, p_2, \ldots, p_{i1})$ and $P_{i2} = (p_1, p_2, \ldots, p_{i2})$ where $i1 < i2$, any candidate event e on* *position k (i2 < k ≤ l) for prefix P_{i2} must also be a candidate on position k for prefix P_{i1}, where l is the period length.*

Proof. With prefix P_{i2}, an event has to recur at least $\frac{min\_gain}{I(P_{i2})+\sum_{j=i2+1}^{l} f_j}$ times on position k to qualify as a candidate while the minimum required number of repetition with prefix P_{i1} is $min\_rep_1 = \frac{min\_gain}{I(P_{i1})+\sum_{j=i1+1}^{l} f_j} =$

$$\frac{min\_gain}{I(P_{i1})+\sum_{j=i1+1}^{i2} f_j+\sum_{j=i2+1}^{l} f_j} \leq \frac{min\_gain}{I(P_{i1})+\sum_{j=i1+1}^{i2} p_j+\sum_{j=i2+1}^{l} f_j}$$

$$= \frac{min\_gain}{I(P_{i2})+\sum_{j=i2+1}^{l} f_j} = \frac{min\_gain}{I(P_{i2})+\sum_{j=i2+1}^{l} f_j}. \text{ Therefore, any candidate on position}$$

k for prefix P_{i2} must be a candidate on position k for prefix P_{i1}. $\square$

Remark 4.3 states that, as the prefix grows longer by filling some unspecified positions, the candidate set of each remaining unspecified position will only shrink monotonically. This provides us with the opportunity to mine surprising patterns by only proceeding with candidate event for each position. Powered by this pruning technique, we develop a progressive approach by starting from extensible prefixes that contain only one filled position (the remaining positions are unspecified) and then proceeding to extensible prefixes with more filled positions gradually to achieve the maximum pruning effect. A candidate event list for each open (i.e., unspecified) position is maintained and continuously refined when more positions are filled. This process continues until all surprising patterns have been identified by examining and extending extensible prefixes. A depth first algorithm is then developed to generate all qualified patterns in a recursive manner.

Another observation we made is that a segment shall not support a pattern P, if it does not support one of P's prefixes. To expedite the process, when we examine a prefix Q, we may screen out those segments that do not support Q and only retain the projected subsequence of Q so that the evaluation of any prefix containing Q would not have to resort to the original sequence. Note that the projected subsequence will also be further refined every time the algorithm proceeds to a prefix with more specified positions. For a given period l, starting with a pattern frame of l slots (without any specific event assigned), potential patterns (or prefixes of potential patterns) are generated progressively by subsequently assigning a candidate event to a yet-unspecified position one at a time. Such assignment may lead to both a refinement of event candidates for the remaining position(s) by applying the above property and a further projection of the projected subsequence onto the remaining open positions. There certainly exist many different orders according to which the open positions are examined. For brevity, we assume the left-to-right order if not specified otherwise in the following discussion. Some heuristic on determining the optimal order is discussed in the optimization section and its effect is shown in experimental results section.

Main Routine

The main procedure of mining statistically significant patterns for a given pattern period is described in the procedure *InfoMiner*. *InfoMiner* is a recursive function. At the kth level of recursion, the patterns with k non-eternal events are examined. For example, all singular patterns (e.g., $(a_1, *, *, *)$) are examined at the initial invocation of *InfoMiner*; at the second level of invocations of *InfoMiner*, all candidate patterns with two non-eternal events (e.g., $(a_1, *, *, a_5)$) are evaluated; an so on. This is achieved by extending the extensible prefixes to include an additional event during each invocation of *InfoMiner* (Line 5-6) and passing the new prefixes to the next level of recursion (Line 8-10). Notice that at most l levels of recursion may be invoked to mine patterns of period l.

Being more specific, at each level of the recursion, we evaluate patterns with certain prefixes in a projected subsequence S. Starting from a null prefix and the sequence in Figure 4.1, a step-by-step example is shown in Figures 4.2, 4.3, 4.4, and 4.5 for prefix *null*, (a_1), (a_1, a_3), and (a_1, a_3, a_4), respectively. First, in the procedure $Repetition\_Calculation(l, S)$ (Line 1), the number of repetitions for each candidate event of each open position is collected from the projected subsequence S. Then the bounded information gain property is employed to refine the candidate list for each remaining open position (Line 2). Finally, for each open position i and each event e in the refined candidate list, a new prefix is created by extending the original one to include the event e on position i (Line 5). Note that this newly created prefix is guaranteed to be extensible and would have the same number of repetitions as the event e at position i (i.e., $repetition[i][e]$). A candidate pattern $P = (prefix, \overbrace{* \cdots *}^{l-i})$ is constructed by filling all remaining open positions following the new_prefix with the eternal event * (Line 6). A subroutine $Pattern\_Validation$ (Line 7) is then invoked to verify whether P has sufficient information gain. The projected subsequence on each new prefix is also generated (Line 8).

InfoMiner($min\_gain, l, S, prefix$)
/* Generate qualified patterns prefixed by $prefix$ and having l other undecided positions from the projected subsequence S where min_gain is the required minimum information gain of a qualified pattern. */
{
1:Repetition_Calculation(l, S)
2:Bounded_Information_Pruning($min\_gain, l, S, prefix$)
3:**for** $i \leftarrow 1$ **to** l **do**
4:    **for each** candidate event e at the ith open position **do**
5:        $new\_prefix \leftarrow (prefix, \overbrace{* \cdots *}^{i-1}, e)$
6:        $P \leftarrow (newprefix, \overbrace{* \cdots *}^{l-i})$

7: Pattern_Validation($min\_gain, P, repetition[i][e]$)
8: $S' \leftarrow$ Projected_Subsequence(S, new_prefix)
9: **if** $S' \neq \emptyset$
10: **then** InfoMiner($min\_gain, l - i, S', new\_prefix$)
}

Repetition_Calculation(l, S)
/* For each of these l undecided positions, compute the number of repetitions for each candidate event for this position from the projected subsequence S. */
{
1:**for** $i \leftarrow 1$ **to** l **do**
2: **for each** candidate event e of the ith position **do**
3: calculate $repetition[i][e]$ from S
}

On the initial call of *InfoMiner*, the entire event sequence and *null* are taken as the input sequence S and the pattern prefix, respectively. In addition, the candidate list of each position consists of all events initially (Figure 4.2(a)). For each successive invocation of *InfoMiner*, a non-empty prefix and its corresponding projected subsequence are passed in as input arguments, and for each remaining open position, those retained events after *Bounded_Information_Pruning* in the previous invocation are taken as the candidates for *Repetition_Calculation*. Consider Figure 4.3, where *InfoMiner* is invoked with $min\_gain = 4.5$, a prefix (a_1), and the projected subsequence given in Figure 4.3(a). The retained events for positions 2, 3, and 4 in Figure 4.2(b) are taken as the candidates in Figure 4.3(b) to calculate the repetition in the projected subsequence of prefix (a_1). After the *Repetition_Calculation* subroutine, a further pruning of these candidates is carried out in the *Bounded_Information_Pruning* subroutine. The refined candidate lists are given in Figure 4.3(c). It is obvious that the candidate list of each open position shrinks after each pruning. This observation can also be made by comparing Figures 4.2(b), 4.3(c), 4.4(c), and 4.5(c), and is demonstrated in the experimental results section. From the candidates in Figure 4.3(c), four new prefixes are generated by extending the previous prefix (a_1): (a_1, a_3), ($a_1, *, a_3$), ($a_1, *, a_4$), and ($a_1, *, *, a_5$) in Figure 4.3(d). The same procedure (i.e., *InfoMiner*), if applicable, is performed recursively on the newly generated prefixes and their corresponding projected subsequences. In the following subsections, we will discuss each subroutine separately.

Bounded_Information_Pruning($min\_gain, l, S, prefix$)
/* For each of these l undecided positions, generate a refined list of candidate events. */

{
1:$max\_info \leftarrow$ information of $prefix$
2:**for** $i \leftarrow 1$ **to** l **do** {
3:    $max[i] \leftarrow$ the maximum value of $I(e)$
4:           for all candidate event e of the ith position
5:    $max\_info \leftarrow max\_info + max[i]$ }
6:$min\_rep \leftarrow \lceil \frac{min\_gain}{max\_info} \rceil$
7:**for** $i \leftarrow 1$ **to** l **do**
8:    remove all events e whose $repetition[i][e] < min\_rep$
9:            from the candidates list
}

Pattern_Validation($min\_gain, P, repetition$)
/* Verify whether a given candidate pattern has sufficient information gain.
If so, add it to the *Result*. */
{
1:$info \leftarrow$ information of P
2:$G(P) = repetition \times info$
3:**if** $G(P) \geq min\_gain$
4:**then** $Result \leftarrow Result \cup \{\langle P, G(P) \rangle\}$
}

Repetition Calculation

The subroutine *Repetition_Calculation* is responsible for collecting the number of repetitions for each candidate event of each position. At the first time this subroutine is invoked, the *prefix* is specified as *null* and every event is considered as a candidate for every position as illustrated in Figure 4.2(a). The entire sequence is scanned to collect the repetition for each event-position combination. The repetitions shown in Figure 4.2(a) is computed from the event sequence in Figure 4.1. During each subsequent invocation, a newly generated non-empty *prefix* is specified and only the projected subsequence on this prefix is passed in as shown in Figure 4.3(a), 4.4(a), and 4.5(a). Only the repetition of each retained candidate event (after pruning in preceding recursions) of each open position in the projected subsequence is collected (e.g., Figure 4.3(b), 4.4(b), and 4.5(b)).

Projected_Subsequence($S, prefix$)
/* Construct the projected subsequence of a data sequence S for a given
prefix. */
{
1:$S' \leftarrow \emptyset$

2:**for each** segment s in S **do**
3:     **if** s supports $(prefix, * \cdots *)$
4:         **then** append s to the end of S'
5:**return** S'
}

Bounded Information Gain Pruning

The procedure $Bounded\_Information\_Pruning$ aims to refine the candidate list for each open position. The general idea is that, by calculating the maximum information (denoted by max_info) that may be carried by a candidate pattern[4] (with a certain prefix) (Line1-4), we can obtain the minimum repetitions (denoted by min_rep) that is necessary to accumulate enough information gain (Line 5). Those events that do not have sufficient repetitions are then removed from the candidate list (Line 6-7). To calculate the value of max_info, for each open position, the event of the maximum information among all candidates for that position is identified (Line 3). The aggregation is taken on the information of each identified event and the information carried by the $prefix$ (Line 1, 4). The value of this aggregation serves as max_info, and min_rep can be calculated by $min\_rep = \lceil \frac{min\_gain}{max\_info} \rceil$. For example, events a_1, a_6, a_4, and a_5 are the events that have the maximum information among the candidates for position 1, 2, 3, and 4 in Figure 4.2(a), respectively. (The information associated with each event is shown in Table 4.2.) Then the maximum information that a qualified pattern may carry can be computed as $max\_info = I(a_1) + I(a_6) + I(a_4) + I(a_5) = 1.06 + 1.45 + 1.16 + 1.06 = 4.73$. In turn, the required minimum repetition is $min\_rep = \lceil \frac{4.5}{4.73} \rceil = 1$ if $min\_gain = 4.5$. Finally, after removing all events whose repetition is less than min_rep, the refined candidate list for each position is shown in Figure 4.2(b).

In the case that the prefix contains some non-eternal event (e.g., $prefix = (a_1)$ in Figure 4.3), the information of the prefix (a_1) is aggregated together with the maximum information from each open position (i.e., the information of a_4, a_4, and a_5, respectively). We have $max\_info = I(a_1) + I(a_4) + I(a_4) + I(a_5) = 1.06 + 1.16 + 1.16 + 1.06 = 4.44$ and $min\_rep = \lceil \frac{4.5}{4.44} \rceil = 2$ if $min\_gain = 4.5$. Figure 4.3(c) shows the refined candidate lists.

The bounded information gain pruning serves as the core of the entire algorithm in the sense that it plays a dominant role in the overall efficiency of the scheme. For each open position, only events that may generate extensible prefixes are retained. Even though the number of candidate events for each position is $O(|E|)$ theoretically, the bounded information gain pruning can substantially reduce the candidate scope in practice via successive candidate refinements. This can be easily seen through the above example and is further verified by the experimental results in the experimental results section. This unique prun-

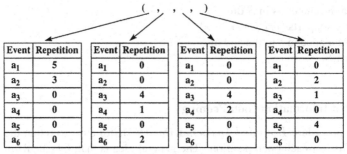

Event	Repetition		Event	Repetition		Event	Repetition		Event	Repetition
a_1	5		a_1	0		a_1	0		a_1	0
a_2	3		a_2	0		a_2	0		a_2	2
a_3	0		a_3	4		a_3	4		a_3	1
a_4	0		a_4	1		a_4	2		a_4	0
a_5	0		a_5	0		a_5	0		a_5	4
a_6	0		a_6	2		a_6	0		a_6	0

(a) initial candidates for each position and their repetitions

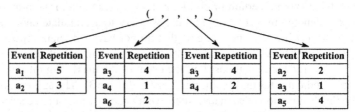

Event	Repetition		Event	Repetition		Event	Repetition		Event	Repetition
a_1	5		a_3	4		a_3	4		a_2	2
a_2	3		a_4	1		a_4	2		a_3	1
			a_6	2					a_5	4

(b) refined candidates after subadditivity pruning where required min_rep = 1

(a_1)	(* , a_3)	(* , * , a_3)	(* , * , * , a_2)
(a_2)	(* , a_4)	(* , * , a_4)	(* , * , * , a_3)
	(* , a_6)		(* , * , * , a_5)

(c) generated new prefixes

Pattern	Information Gain
(a_1 , * , * , *)	5.30

(d) qualified patterns with min_gain = 4.5

Figure 4.2.   Initial Invocation of *InfoMiner*

ing technique distinguishes our proposed method from any existing algorithm designed for mining frequent patterns.

Pattern Validation

The subroutine $Pattern\_Validation$ is used to validate a candidate pattern P. The information of P can be easily obtained by aggregating the information of the event on each position. A decision is then made on whether the information gain (i.e., $repetition \times I(P)$) meets the requirement. In the case that P carries sufficient information gain, P is inserted into $Result$. In Figure 4.3(e), $(a_1, a_3, *, *)$ and $(a_1, *, *, a_5)$ are added into $Result$.

$a_1\ a_3\ a_4\ a_5 \mid a_1\ a_4\ a_3\ a_3 \mid a_1\ a_4\ a_3\ a_3 \mid a_1\ a_3\ a_3\ a_5 \mid a_1\ a_3\ a_4\ a_5 \mid a_1\ a_3\ a_4\ a_5$

(a) projected subsequence of prefix (a $_1$)

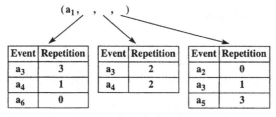

(b) repetitions of each candidte in the projected subsequence

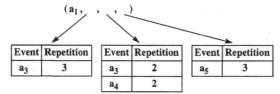

(c) refined candidates by subadditivity property where min_rep =

(a_1, a_3) $(a_1, *, a_3)$ $(a_1, *, *, a_5)$
            $(a_1, *, a_4)$

(d) generated new prefixes

Pattern	Information Gain
$(a_1, a_3, *, *)$	5.19
$(a_1, *, *, a_5)$	6.36

(e) qualified pattern with min_gain = 4.5

Figure 4.3.   Level 2 Invocation of *InfoMiner* for prefix (a_1)

We note that, even if the pattern P generated from some *newprefix* (Line 5-6) in procedure *InfoMiner* does not carry sufficient information gain, the *newprefix* may still be extended to a pattern that satisfies the information gain threshold (i.e., *newprefix* is extensible). For example, the pattern $(a_1, *, a_4, *)$ (generated from prefix $(a_1, *, a_4)$ in Figure 4.3(d)) has information gain 4.44 (which is less than *min_gain*). However, the prefix $(a_1, *, a_4)$ can be extended to $(a_1, *, a_4, a_5)$ whose information gain is greater than *min_gain*. This illustrates the essential difference between the Apriori property and the bounded information gain property.

a_1 a_3 a_4 a_5 a_1 a_3 a_3 a_5 a_1 a_3 a_4 a_5 a_1 a_3 a_4 a_5

(a) projected subsequence of prefix (a_1, a_3)

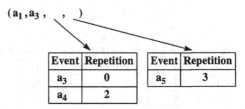

Event	Repetition		Event	Repetition
a_3	0		a_5	3
a_4	2			

(b) repetitions of each candidte in the projected subsequence

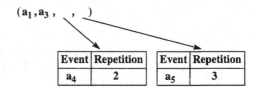

Event	Repetition		Event	Repetition
a_4	2		a_5	3

(c) refined candidates by subadditivity property where min_rep

(a_1, a_3, a_4) $(a_1, a_3, *, a_5)$

(d) generated new prefixes

Pattern	Information Gain
$(a_1, a_3, a_4, *)$	5.78
$(a_1, a_3, *, a_5)$	8.37

(e) qualified patterns with min_gain = 4.5

Figure 4.4. Level 3 Invocations of *InfoMiner* for prefixes (a_1, a_3)

Projected Subsequence Construction

The subroutine $Projected\_Subsequence$ is very straightforward. Given a prefix, the input sequence S is scanned and the segments are selected to form the projected subsequence. Note that all segments in this projected subsequence share a common prefix. For example, the projected subsequence for prefix (a_1) is given in Figure 4.3(a) and that for prefix (a_1, a_3) is given in Figure 4.4(a). We shall see in the optimization section that this characteristic provides the opportunity for further optimization.

$a_1\ a_3\ a_4\ a_5\ a_1\ a_3\ a_4\ a_5\ a_1\ a_3\ a_4\ a_5$

(a) projected subsequence of prefix (a_1, a_3, a_4)

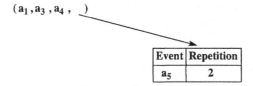

(b) repetitions of each candidte in the projected subsequence

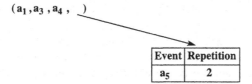

(c) refined candidates by subadditivity property where min_rep

$(a_1,\ a_3,\ a_4, a_5)$

(d) generated new prefixes

Pattern	Information Gain
(a_1, a_3, a_4, a_5)	7.90

(e) qualified patterns with min_gain = 4.5

Figure 4.5.   Level 4 Invocations of *InfoMiner* for (a_1, a_3, a_4)

Discussion: Finding the K Most Surprising Patterns

In the previous section, we presented an algorithm to find patterns whose information gain is greater than a threshold min_gain. In some applications, users may want to specify the number of most surprising patterns to be returned instead. In this section, we investigate the problem of mining the K most surprising patterns. In fact, this can be achieved with the following minor modification to the algorithm presented in the previous section. The set $Result$ is maintained throughout the entire process to hold the set of K most surprising patterns discovered so far, and is initialized to be null. The threshold min_gain is always equal to the minimal information gain carried by any pattern in $Result$ and is set to zero at the beginning. Each time after discovering a new pattern P

whose information gain $G(P)$ is above the current threshold min_gain, P is added to $Result$ and, in the case where $Result$ already contains K patterns, the pattern with the smallest information gain in $Result$ will be replaced. The value of min_gain is then adjusted to the minimal information gain of any pattern in $Result$. Clearly, with the algorithm proceeds, min_gain increases, thus, min_rep (used in the algorithm) also increases, and less candidates remain. At the end, the set $Result$ contains the K most surprising patterns (i.e., highest information gain).

In the experimental results section, we show the performance of finding the K most surprising patterns with comparison to finding these K patterns when the appropriate information gain threshold is given. When K is relatively small (e.g., 10), finding the K most surprising patterns may take twice as much time as finding patterns with given information gain threshold. On the other hand, when K is large (e.g., 1000), the disadvantage diminishes (e.g., less than 5%). We will discuss it in more detail in experimental results section. Since the absolute response time of finding a small number of patterns that meet a very high information gain threshold is relatively short, the prolonged response time is still tolerable in most cases.

Optimization Issue

In this section, we explore some further optimization technique that may be used to improve the performance of *InfoMiner*. It is obvious that, as the algorithm proceeds from patterns with single filled position to more complex ones, we may examine open positions in different orders. Certainly, the imposed order would affect the performance of *InfoMiner* in the sense that, at a given level of the recursion, both the length of projected subsequence and the number of candidate events for an open position would be different. We now explore this issue in detail and propose some heuristic on how to choose an optimal order.

Given a pattern prefix p, let S_p be the projected subsequence of p. In order to generate qualified patterns with prefix p, *InfoMiner* scans S_p to collect the repetition of each candidate event at each open position and simultaneously compute the projected subsequences of some newly constructed prefixes. Therefore, the length of S_p is an important factor towards the performance of *InfoMiner*. In general, the length of a projected subsequence is a function of the length of the corresponding prefix, which decreases as the prefix includes more filled positions. This can be easily observed by examining the projected subsequences of (a_1), (a_1, a_3), and (a_1, a_3, a_4) in Figure 4.3(a), 4.4(a), and 4.5(a), respectively. It would be desirable if we can minimize the length of the projected subsequence for any given prefix length. To achieve this goal, we propose a heuristic, namely *shortest projection first* (SPF), to determine the order to examine the open positions. The basic idea is that, the length of a projected subsequence that is going to be constructed for a newly generated prefix (in an invocation of *InfoMiner*) can be estimated from the repetition of the corresponding candidate event. For instance, the repetition of a_1 at the first position is 5 in Figure 4.2 and the length of the projected subsequence for (a_1) is going to contain $(5 + 1) \times 4 = 24$ events. Let $o_1, o_2, \ldots, o_l$ be the set of open positions and $length_i$ be the total length of projected subsequences for the prefix generated by filling position o_i with some candidate event. Then, all open positions can be ordered in ascending order of $length_i$. Let's consider the example in Figure 4.2. If the first position is specified, the total length of all projected subsequences is $(5 + 1) \times 4 + (3 + 1) \times 4 = 40$. On the other hand, if we consider the third position first, then the total length of all resulting projected subsequences is $(4 + 1) \times 4 + (2 + 1) \times 4 = 32$. Thus the third position should be examined before the first position according to this order. As a result, the total length of projected subsequences is minimized. This step can be performed after the $Bounded\_Information\_Pruning()$ to provide an order among all open positions. We will show in the experimental section that this optimization can produce as much as 40% reduction towards the length of projected subsequences.

Experimental Results for InfoMiner

We implemented InfoMiner in C programming language on an IBM RS-6000 (300 MHz CPU) with 128MB running AIX operating system. To analyze the benefits of the information model and the performance of the InfoMiner algorithm, we employ two real traces and four synthetic traces.

IBM Intranet and Scour Traces

The IBM Intranet traces consist of 160 critical nodes, e.g., file servers, routers, etc., in the IBM T. J. Watson Intranet. Each node issues a message in response of certain situation, e.g., CPU saturation, router interface down, etc. There are total 20 types of messages. We treat a certain message from a particular node as a distinct event, thus there are total 500 distinct events in the trace because a certain type of node may only send out 4 or 5 types of messages. The IBM Intranet trace consists of 10,000 occurrences of the events. With the support threshold, the 100 most frequent patterns all contain the "I am alive" message which is a control message that a network node sends out periodically, e.g., every minute. Despite its high frequency, it is not particularly useful. On the other hand, by applying InfoMiner on this trace, we found some surprising patterns that are also interesting. For example, the pattern $(node_a\_fail, *, node_b\_saturated, *)$ has the eighth highest information gain. This pattern means that, shortly after a router $(node_a)$ fails, the CPU on another node $(node_b)$ is saturated. Under a thorough investigation, we found that $node_b$ is a file server and after $node_a$ fails, all requests to some files are sent to $node_b$, which cause the bottleneck. This particular pattern is not even ranked within the top 1000 most frequent patterns under the support model.

Scour is a web search engine that is specialized for multimedia contents. The web URL of the Scour search engine is "http://www.scour.net". The scour servers consist of a cluster of machines. Every five minutes, a distributed sensor records the CPU utilization of every node and the average is taken as the global CPU utilization level. We discretize the global CPU utilization level into events: event A stands for the utilization level between 0 and 0.05, B stands for the utilization level between 0.05 and 0.1, and so on. The scour trace consists of 28800 occurrences of 19 events (100 days). The event corresponding to the utilization between 0.95 and 1 does not occur in the trace.

Since the utilization level of a node is typically around 0.3, most of the frequent patterns consists of symbol F or G, such as $(F, G, *)$ or $(G, *, F)$, and so on. However, these patterns are not interesting nor useful. On the other hand, with the information gain measure, we are able to find some interesting and useful patterns, such as a pattern that S (utilization [0.9, 0.95]) followed by Q (utilization [0.8, 0.85]), which, in turn, is followed by P (utilization [0.75, 0.8]). This pattern means that if the nodes are saturated and the response time increases significantly, then many users may stop access the site shortly after, thus the workload decreases immediately. Although the support of this pattern is low, the information of this pattern is quite high (ranked as the 8th highest) due to the rare occurrences of S and Q. In the support model, this pattern is ranked beyond 10,000th. This means that we have to examine more than 10,000 other patterns before discovering this pattern in the support model. (The complete comparison of the number of patterns found in the information model and the support model for both IBM trace and Scour trace are shown in Table 4.1 above.)

Table 4.3 shows the number of discovered patterns and the response time under the information model and support model in the scour trace. In this table, the minimum support threshold (min_sup) is set based on the min_gain value so that the patterns found in the support model would include all patterns found in the information model. In addition to the set of patterns returned by the information model, the support model also discovers a huge number of non-surprising patterns. This not only creates an encumbrance to the user but also leads to a slower response time. It is easy to see that the response time of the support model is an order of magnitude slower than that of the information model. In addition, we find that the bounded information gain pruning is very effective in the sense that more than 95% of the patterns are pruned in this experiment.

To further compare the surprising patterns and frequent patterns, we train two classifiers, one based on top 1000 frequent patterns and the other based on top 1000 surprising patterns, to predict whether the system operates normally

Table 4.3. Number of Discovered Patterns and Response Time for the Scour Trace

Information Model		Support Model	
min_gain	Num. of Patterns	min_sup	Num. of Patterns
$-\log_{19} 0.00001 = 3.91$	8	0.000212	16,953
$-\log_{19} 0.00003 = 3.54$	21	0.000208	17,021
$-\log_{19} 0.0001 = 3.13$	53	0.000201	17,605
min_gain	Response Time	min_sup	Response Time
$-\log_{19} 0.00001 = 3.91$	1.5 min	0.000212	25.2 min
$-\log_{19} 0.00003 = 3.54$	2.7 min	0.000208	27.1 min
$-\log_{19} 0.0001 = 3.13$	4.2 min	0.000201	28.7 min

at any given time. Each pattern is treated as a feature. The true status of the system (normal versus abnormal) is given by the administrator. The support vector machine (SVM) is adopted as the classification model. We found that SVM trained on frequent patterns yields 34.8% adjusted accuracy[5] whereas the classifier build upon surprising pattern achieve an adjusted accuracy of 77.3%. This further demonstrates the superiority of surprising patterns.

Synthetic Sequences

To analyze the performance of the InfoMiner Algorithm, four sequences are synthetically generated. Each sequence consists of 1024 distinct events and 20M occurrences of events. The synthetic sequence is generated as follows. First, at the beginning of the sequence, the period length l of the next pattern is determined, which is geometrical distributed with mean μ_l. The number of events in a pattern is randomly chosen between 1 and l. The number of repetitions m of this pattern is geometrical distributed with mean μ_m. The events in the pattern are chosen according to a normal distribution. (This means that some events occurs more frequently than other.) However, the pattern may not perfectly repeat itself for m times. To simulate the imperfectness of the subsequence, we employ a parameter δ to control the noise. δ is uniformly distributed between 0.5 and 1. With probability δ, the next l events match the pattern. Otherwise, the next l events do not support the pattern. The replacement events are chosen from the event set with the same normal distribution. This subsequence ends when there are m matches, and a new subsequence for a new pattern starts. This process repeats until it reaches the end of the sequence. Four sequences are generated based on values of μ_l and μ_m in Table 4.4.

Standard InfoMiner

In this subsection, we analyze the InfoMiner algorithm without the optimization technique presented in optimization section. The main pruning power of the InfoMiner algorithm is provided by the bounded information gain pruning

Table 4.4.   Parameters of Synthetic Data Sets

Data Set	μ_l	μ_m	Distinct events	Total Events
$l3m20$	3	20	1024	20M
$l100m20$	100	20	1024	20M
$l3m1000$	3	1000	1024	20M
$l100m1000$	100	1000	1024	20M

technique. In our InfoMiner algorithm, for each prefix, we prune the candidate events on each remaining open position. Although the number of candidate events on each open position can be $|E|$ theoretically, in practice the average number of candidates in each open position decreases dramatically with the increase of the number of specified positions (i.e., the length of the prefix). This is due to the fact that the value of max_info is estimated from the candidate event with the highest information on each open position. Thus, with more positions specified, the max_info value decreases. In turn, the min_rep threshold ($min\_rep = \lceil \frac{min\_gain}{max\_info} \rceil$) increases and more candidate events are pruned. We conduct experiments with our InfoMiner algorithm on the four synthetic sequences. Figure 4.6(a) shows the average number of remaining candidate events on each open position as a function of the number of specified positions (i.e., the length of the prefix). The number of candidate events decreases dramatically with the number of specified positions. With data set $l3m20$ and $l3m1000$, since the average pattern length is 3, there is no candidate after 5 or 6 positions are specified. In addition, with all four data sets, when the number of specified positions is greater than 3, the average number of events on each open position is very small (i.e., less than 0.4). This leads to the overall efficiency of the InfoMiner algorithm.

The overall response time of the InfoMiner algorithm largely depends on the min_gain threshold. We ran several tests on the above data sets with different min_gain thresholds. Figure 4.6(b) shows the response time for each data set. Our bounded information gain pruning can reduce a large number of patterns. More than 99% of the patterns are pruned. When the min_gain threshold increases, the pruning effect becomes more dominant because more patterns can be eliminated by the bounded information gain pruning. Thus, the response time improves with increasing min_gain threshold on all four data sets. (Note that the Y-axis is in log scale in Figure 4.6(a) and (b).) To explore the scalability of the InfoMiner algorithm, we also experiment with event sequences of different lengths, from 1 million to 100 million. We found that the response time of InfoMiner is linear with respect to both the length of the event sequence and the period length.

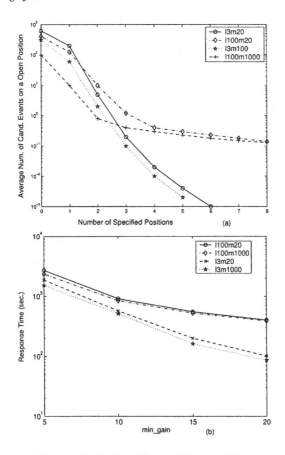

Figure 4.6. Pruning Effects and Response Time

Optimization Technique

In this subsection, we analyze the improvement rendered by the optimization techniques presented in the optimization section. The shortest projection first (SPF) technique reduces the average length of a projected subsequence. Figure 4.7(a) shows the improvement provided by the SPF technique on the synthetic sequences. Since all four synthetic sequences show a similar rate of improvement, we use the $l100m20$ sequence as the representative in Figure 4.7(a). The improvement is measured as $\frac{L-L_{spf}}{L}$ where L and L_{spf} are the average length of a projected subsequence with the standard InfoMiner and with the SPF technique, respectively. It is evident that the most (accumulative) improvement is achieved when the number of specified positions is 3. This is due to the fact that when the projected subsequence is substantially long and the

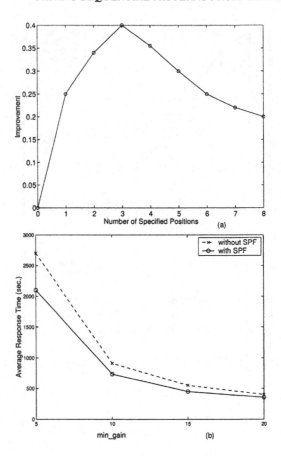

Figure 4.7.   Improvements of the SPF Heuristics

candidate event list of each open position is comparatively large, there exists a relatively large room for further improvement, while such potential diminishes as the number of specified positions in the prefix increases. Figure 4.7(b) illustrates the response time of the InfoMiner algorithm with and without the SPF heuristic. It is obvious that the SPF technique can achieve 20% improvement on the overall response time of our InfoMiner algorithm.

The K Most Surprising Patterns

Let $info\_gain(K)$ be the Kth highest information gain among all patterns. More specifically, $info\_gain(K)$ is a monotonic decreasing function of K. Figure 4.8(a) shows the ratio of the response time of finding the K most surprising patterns (without knowing the value of $info\_gain(K)$ in advance) and

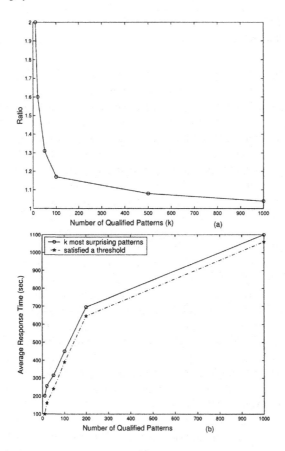

Figure 4.8. Performance of mining the K most surprising patterns

finding patterns above the given threshold $info\_gain(K)$. Obviously, these two settings would produce exactly the same result. The x-axis shows the number of desired patterns, K. The higher the value of K, the less the overhead. (The overhead is less than 5% when $K = 1000$.) When K is small, finding the K most surprising patterns is about twice as long as that of find these K patterns when $info\_gain(K)$ is given. This is due to the fact that the threshold $info\_gain(K)$ is relatively high when K is small. Consequently, the overhead is higher since the mining process takes a relatively longer time to raise min_gain (from zero) to the proper threshold $info\_gain(K)$.

Figure 4.8 (b) illustrates the absolute response time of these two settings. The smaller the value of K, the shorter the overall response time. This is largely due to the effect of rapid adjustment of the threshold min_gain towards the value of $info\_gain(K)$ so that a large amount of patterns can be quickly pruned at early stage of the mining process.

2.   Stamp

We have presented the InfoMiner algorithm, which uses the information gain to measure the importance/significance of the occurrence of a pattern. The information gain of an occurrence of a rare event is high while the information gain of the occurrence of a frequent event is low. Thus, we are able to find the statistically significant patterns with the information gain threshold. However, the major limitation of this model is that it does not take into account the location of the occurrences of the patterns in the sequence. Let's take a look at two sequences: $S_1 = a_1, a_2, a_3, a_3, a_2, a_3, a_1, a_2, a_4, a_5, a_1, a_2$ and $S_2 = a_1, a_2, a_1, a_2, a_1, a_2, a_3, a_3, a_2, a_3, a_4, a_5$. The elements in the two sequences are identical. The only difference is the order of the events. The pattern (a_1, a_2) repeats perfectly in the first half of S_2 while it scatters in S_1. The two patterns have the same information gain in the two sequences. In some applications (e.g., repeats discovery in bio-informatics domain), a series of consecutive repeats are considered more significant than the scattered ones. That is, there should be some "penalty" associated with the gap between pattern repeats. As a result, the measure of generalized information gain (GIG) was introduced [19] to capture the significance of a pattern in a sequence/subsequence. The occurrence of a pattern will be given a positive GIG while a mis-occurrence (or a gap) will generate a negative GIG. The overall generalized information gain will be the aggregate GIG of all occurrences and mis-occurrences of the pattern in a sequence/subsequence.

Since the characteristics of a sequence may change over time, many patterns may only be valid for a period of time. The degree of significance (i.e., generalized information gain) of a pattern may be diluted if we only consider the entire event sequence. In addition, a user may be interested not only in a significant pattern, but also where/when the pattern is significant as well. The identification of significant pattern in a subsequence is of great importance in many applications. In our proposed scheme, a user can specify the minimum GIG that a significant pattern must carry over a subsequence of data. Upon satisfying this requirement, the subsequence(s) that maximizes the GIG of a pattern will be identified. In the previous example, the pattern (a_1, a_2) is very significant in the first half of S_2, but may not be significant over the entire sequence.

Although the generalized information gain is a more meaningful metric for the problems addressed previously, it does not preserve the *downward closure* property (as the *support* does). For example, the pattern $(a_1, a_2, *)$ may have sufficient GIG while both $(a_1, *, *)$ and $(*, a_2, *)$ do not. We can not take advantages of the standard pruning technique developed for mining association rules. The observation that the *triangle inequality*[6] is still preserved by the generalized information gain motivates us to devise a threefold algorithm as the core of our pattern discovery tool, STAMP.

1 First, the significant patterns involving one event are discovered. Two novel strategies, **optional information surplus pruning** and **maximum information gain counting**, are proposed to dramatically reduce the search space.

2 Next, candidate patterns involving multiple events are generated based on the *triangle inequality*.

3 All candidate patterns are validated and for each pattern which is significant, the corresponding subsequence containing the pattern is also identified.

Generalized Information Gain

In this section, we provide a brief overview of the model of generalized information gain [19] and discuss its properties. Let $\Im = \{a_1, a_2, \ldots, \}$ be a set of events and D be a sequence of events in $\Im$.

Definition A **pattern** with **period** l is an array of l events $(p_1, p_2, \ldots, p_l)$, each of which is either an event in $\Im$ or *, i.e., $p_j \in \Im \cup * \ (1 \leq j \leq l)$. We say that the jth position is **instantiated** if $p_j \in \Im$. For any two patterns $P = (p_1, p_2, \ldots, p_l)$ and $P' = (p'_1, p'_2, \ldots, p'_l)$ of the same period l, P is a **superpattern** of P' (P' is a **subpattern** of P) if $p'_j = *$ or $p_j = p'_j$, for all $1 \leq j \leq l$.

Note that an event can appear at multiple positions in a pattern. For example, $(a_1, a_4, *, *, a_4)$ is a pattern of period 5 and its first, second and fifth positions are instantiated. It is also a superpattern of $(a_1, *, *, *, a_4)$.

Definition For an event $a_j \in \Im$ and a sequence D of N events, let $Prob(a_j)$ be the expected probability that a_j occurs at any given position in D[7]. Then the **information** of a_j with respect to D is defined as $I(a_j) = \log \frac{1}{Prob(a_j)} = -\log Prob(a_j)$. The information of the "eternal" event * is always 0[8].

In practice, the probability of each event can be determined in many ways as described in the InfoMiner section. We restate these options here.

- $Prob(a_1) = Prob(a_2) = \cdots = Prob(a_j) = \cdots = \frac{1}{|\Im|}$;

Event	Machine 1	Machine 2	Machine 3	Machine 4	Machine 5	Probability	Information
a1	low	low	low	low	low	0.0010	0.997
a2	low	low	low	low	relatively low	0.0024	0.87
a3	low	low	low	low	relatively high	0.0013	0.959
a4	low	low	low	low	high	0.0001	1.33
a5	low	low	low	relatively low	low	0.0018	0.912
a6	low	low	low	relatively low	relatively low	0.0008	1.03
a7	low	low	low	relatively low	relatively high	0.0005	1.1
a8	low	low	low	relatively low	high	0.0023	0.876
a9	low	low	relatively low	low	low	0.0007	1.05
.	.	.	.	.	.	.	.
.	.	.	.	.	.	.	.
.	.	.	.	.	.	.	.
a1024	high	high	high	high	high	0.0003	1.17

Figure 4.9. The Set of Possible Events

- $Prob(a_j) = \frac{Num_D(a_j)}{N}$ for all $a_j \in \Im$ where $Num_D(a_j)$ and N are the number of occurrences of the event a_j in D and the length of the event sequence D, respectively;

- $Prob(a_j)$ is determined according to some domain knowledge.

In this chapter, we adopt the second option and will not assume the same probability for every event, i.e. an occurrence of frequent event carries less information than a rare event. Note that this also coincides with the original intention of *information* in the data communication community. **Definition**

The **information** of a pattern $P = (p_1, p_2, \ldots, p_l)$ is the summation of the information carried by each individual position, i.e., $I(P) = \sum_{1 \leq j \leq l} I(p_j)$.

Figure 4.9 shows the set of events corresponding to different workload states of a web server. There are total 1024 different events. Their probabilities of occurrence are arbitrarily assigned and the corresponding information is calculated accordingly. We use 1024^9 as the base in the calculation. The information of pattern $(a_1, a_4, *, *, a_4)$ is $I(a_1)+I(a_4)+I(*)+I(*)+I(a_4) = 0.997+1.33+0+0+1.33 = 3.657$. After defining the information of a pattern, now we begin to formulate the definition of information gain of a pattern in a subsequence of events.

Given a pattern $P = (p_1, p_2, \ldots, p_l)$ with period l and a sequence of l events $D' = d_1, d_2, \ldots, d_l$, we say that D' is **in compliance with** P at position j $(1 \leq j \leq l)$ iff either $p_j = *$ or $p_j = d_j$ holds. For example, the sequence a_1, a_1, a_2, a_3 is in compliance with the pattern $(a_1, a_2, a_3, *)$ at positions 1 and 4.

Given a pattern $P = (p_1, p_2, \ldots, p_l)$ with period l and a sequence of l events $D' = d_1, d_2, \ldots, d_l$, we say that P **matches** D' (or D' **supports** P), iff D' is in compliance with P at every position $j(1 \leq j \leq l)$. For instance, the sequence a_1, a_4, a_2, a_3, a_4 supports the pattern $(a_1, a_4, *, *, a_4)$ while the sequence a_1, a_4, a_2, a_3, a_6 does not support it since the sequence is not in compliance with the pattern on the last position.

Definition Given a pattern P with period l and a sequence D of $N(N \geq l)$ events: $d_1, d_2, \ldots, d_N$, the **support** of P within D is the number of subsequences $d_{l \times j+1}, d_{l \times j+2}, \ldots, d_{l \times j+l}$ that match P.

Intuitively, the event sequence can be viewed as a list of segments, each of which consists of l contiguous events. There would be $\lfloor N/l \rfloor$ full segments, among which the segment that P matches will count for the support of P. The "*" symbol is a wild card which matches any symbol.

Consider two subsequences $D_1 = a_1\, a_2\, a_1\, a_2\, a_1\, a_2$ and $D_2 = a_1\, a_2\, a_1\, a_1$ $a_1\, a_2\, a_1\, a_2$ for the pattern $P = (a_1, a_2)$. The support of P in D_1 is the same as that in D_2, which is 3. However, the generalized information gain of P with respect to D_1 should be higher than that of D_2 because there is no noise in D_1 but some noise in D_2. Therefore, D_2 should "pay some penalty" for its noise, i.e., taking away some generalized information gain from D_2. The amount of generalized information gain taken away depends on how D_2 can be repaired to perfection. In this case, if we replace an event a_1 with a_2, then D_2 would be perfect for P. Thus, we decide to take away the information of a_2 (i.e., the information loss of P for the mismatched period in D_2) from D_2. **Definition**

Given a pattern $P = (p_1, p_2, \ldots, p_l)$ with period l and a sequence of l events $D' = d_1, d_2, \ldots, d_l$, the **information loss** of D' on position j with respect to P is the information of the event p_j iff D' is not in compliance with P at position j and there is no information loss otherwise. The overall information loss of D' with respect to P is the summation of the information loss of each position.

For example, the information loss of $a_1\, a_4\, a_5\, a_2\, a_3$ on position 2 with respect to $(a_1, a_4, *, *, a_4)$ is 0 while the infomation loss on position 5 is $I(a_4) = 1.33$.

Definition Given a pattern P with period l and a sequence D of $N(N \geq l)$ events: $d_1, d_2, \ldots, d_N$, the **information loss** of D with respect to P is the summation of the information loss of each segment $d_{l \times j+1}, d_{l \times j+2}, \ldots, d_{l \times j+l}$ with respect to P. The **generalized information gain** of D with respect to P is defined as $I(P) \times (S_D(P) - 1) - L_D(P)$ where $I(P)$, $S_D(P)$, and $L_D(P)$

are the information of P, the support of P within D, and the information loss of D with respect to P, respectively.

In a subsequence, the first match of a pattern is viewed as an example, and only subsequent matches contribute to the generalized information gain[10]. With the generalized information gain concept, let's consider the event sequence a_1 a_1 a_1 a_1 a_1 a_2 a_1 a_1 a_2 a_1 a_3 a_2 a_3 a_1 a_3 a_3 a_1 a_3 a_1 a_1. In this sequence, the information of the occurrence of a_1 is $-\log_3(12/20) = 0.46$ while the information of the occurrence of a_2 is $-\log_3(3/20)) = 1.73$. The generalized information gain of $(a_2, *, *)$ in subsequence $a_2\, a_1\, a_1\, a_2\, a_1\, a_3\, a_2\, a_3\, a_1$ is 3.46 while the generalized information gain of $(a_1, *, *)$ in subsequence $a_1\, a_1\, a_1\, a_1$ $a_1\, a_2\, a_1\, a_1\, a_2\, a_1\, a_3\, a_2$ is 1.38. This is due to the fact that event a_1 occurs much more frequent than a_2, and hence it is likely that $(a_1, *, *)$ also occurs frequently in the event sequence. On the other hand, event a_2 occurs relatively scarce, thus the occurrence of $(a_2, *, *)$ carries more information. Therefore, if $g = 3$ is specified as the generalized information gain threshold, then pattern $(a_2, *, *)$ would satisfy the threshold, but not $(a_1, *, *)$. However, with the traditional support confidence thresholds, $(a_1, *, *)$ always has higher support value than $(a_2, *, *)$.

Definition Given a pattern P, a sequence D and a generalized information gain threshold g, if there exists a subsequence D' of D so that the generalized information gain of D' with respect to P is at least g, then P is a **valid pattern**.

Theoretically, the period of a valid pattern could be arbitrary, i.e., as long as the event sequence. In reality, a user can specify an upperbound of period length according to his/her domain knowledge. As a result, we use L_{max} to denote the maximum period allowed for a pattern. However, L_{max} can be arbitrarily large, e.g., ranging to several thousands. Now we can rephrase our problem model by employing the generalized information gain metric. For a sequence of events D, a generalized information gain threshold g, and a period bound L_{max}, we want to discovery all valid patterns P whose period is less than L_{max}.

For each valid pattern P, we want to find the subsequence which maximizes the generalized information gain of P. In the remainder of this section, we give some more definitions which enable us to present our approach STAMP and communicate to readers more effectively.

Definition For any two patterns $P = (p_1, p_2, \ldots, p_l)$ and $P' = (p'_1, p'_2, \ldots, p'_l)$ of the same period l, P and P' are **complementary** if either $p_j = *$ or $p'_j = *$ for all $1 \leq j \leq l$.

A set of patterns of the same period are said to be *complementary* if every pair of patterns in the set are complementary. Given a set Π of complementary patterns of the same period l, the **minimum common superpattern (MCSP)** of Π is the pattern P of period l, which satisfies the following two conditions.

- Each pattern in Π is a subpattern of P.

- There does not exist a subpattern P' of P $(P' \neq P)$ such that each pattern in Π is also a subpattern of P'.

It follows from the definition that the information of the MCSP of a set, Π, of complementary patterns is the summation of the information of each pattern in Π. For example, $(a_1, a_3, *, *, *)$, $(*, *, *, a_2, *)$, and $(*, *, *, *, a_4)$ are complementary and their MCSP is $(a_1, a_3, *, a_2, a_4)$. The information of $(a_1, a_3, *, a_2, a_4)$ is $I(a_1) + I(a_3) + I(a_2) + I(a_4)$ which is exactly the summation of the information of $(a_1, a_3, *, *, *)$, $(*, *, *, a_2, *)$, and $(*, *, *, *, a_4)$. For a given event segment $D' = d_1, d_2, \ldots, d_l$ and a set, Π, of complementary patterns, the information loss of D' with respect to the MCSP of Π satisfies the following equality

$$L_{D'}(MCSP(\Pi)) = \sum_{P \in \Pi} L_{D'}(P)$$

where $L_D(P)$ is the information loss of D' with respect to P. The rationale is that if D' is not in compliance with a pattern P in Π on position j, then the jth position must be instantiated and D' must not be in compliance with the MCSP of Π on position j either. For instance, the information loss of the segment a_1, a_1, a_2, a_1, a_4 with respect to $(a_1, a_3, *, a_2, a_4)$ is $I(a_3) + I(a_2)$, which is equal to the aggregate of information loss of that segment with respect to $(a_1, a_3, *, *, *)$, $(*, *, *, a_2, *)$, and $(*, *, *, *, a_4)$. In general, for any event sequence D, the overall information loss of D with respect to the MCSP of a set of complementary patterns Π is equal to the summation of the information loss of D with respect to each pattern in Π.

PROPOSITION 2.1 **(Triangle Inequality)** *Given an event sequence D and two complementary patterns P and P' of the same period, let Q be the minimum common super pattern of P and P'. Then the generalized information gain of D with respect to Q is at most the summation of that of P and P'.*

Proof. Since P and P' are complementary, the information of Q is $I(Q) = I(P) + I(P')$ and $L_D(Q) = L_D(P) + L_D(P')$ for any event sequence D. Then for any sequence D, the generalized information gain of D with respect to Q is

$$
\begin{aligned}
& I(Q) \times (S_D(Q) - 1) - L_D(Q) \\
= {} & (I(P) + I(P')) \times (S_D(Q) - 1) - L_D(P) - L_D(P') \\
= {} & I(P) \times (S_D(Q) - 1) - L_D(P) + I(P') \times (S_D(Q) - 1) \\
& -L_D(P') \\
\leq {} & I(P) \times (S_D(P) - 1) - L_D(P) + I(P') \times (S_D(P') - 1) \\
& -L_D(P')
\end{aligned}
$$

because the support of Q in D (i.e., $S_D(Q)$) is at most the support of P in D (i.e., $S_D(P)$). Thus this proposition holds. $\square$

Proposition 2.1 can be easily generalized to a set of complementary patterns, which is stated as follows.

PROPOSITION 2.2 *Given an event sequence D and a set of complementary patterns Π, let Q be the minimum common super pattern of Π, then the generalized information gain of D with respect to Q is at most the summation of that of each pattern in Π.*

Stamp Algorithm

In this section, we outline the general strategy we use to mine patterns that meet certain generalized information gain threshold g. There exist three challenges for mining patterns with information gain: (1) The number of different patterns is

$$
\sum_{0 < l \leq L_{max}} \left(|\Im|^l - 1 \right) = O(|\Im|^{L_{max}})
$$

where $|\Im|$ and L_{max} are the overall number of distinct events and the maximum period length, respectively. Since L_{max} can be quite large, e.g., in the thousands, it is infeasible to verify each pattern against the data directly. Some pruning mechanism has to be developed to circumscribe the search space. (2) By definition, the generalized information gain measure does not have the property of downward closure as the traditional *support* measure does. For the sequence shown in Figure 4.10, the generalized information gains[11] of the pattern $(a_4, *, *)$ and $(*, a_9, *)$ are $(5 - 1) \times 1.33 - 1.33 - 1.33 = 2.66$ and $(6 - 1) \times 1.05 - 1.05 = 4.2$, respectively; while the generalized information gain of $(a_4, a_9, *)$ is $(5 - 1) \times 2.38 - 1.33 - 1.05 - 1.33 = 5.81$, which is greater than that of $(a_4, *, *)$ and $(*, a_9, *)$. If the generalized information gain threshold is set to $g = 5$, then only $(a_4, a_9, *)$ qualifies while the other two do not. This prevents us from borrowing existing algorithms developed for association rule problems to mine the qualified patterns. (3) The subsequence concept introduced in this chapter poses a difficult challenge to determine when a subsequence should start and end. If a pattern misses some "matches", it is

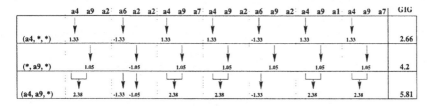

Figure 4.10. Violation of the Downward Closure

hard to tell whether this signals the end of a subsequence or this merely means some noise within a subsequence.

Fortunately, the *triangle inequality* holds for the generalized information gain. In other word, for a set of complementary patterns Π, the generalized information gain of the minimum common super-pattern (MCSP) of Π is always less than or equal to the sum of that of each individual pattern in Π over the same subsequence of events. For example, the generalized information gain of $(a_4, a_9, *)$ is less than the sum of that of $(a_4, *, *)$ and $(*, a_9, *)$ over the same subsequence as demonstrated in Figure 4.10. Inspired by this observation, we can first collect the generalized information gain of all singular patterns, and then generate candidate patterns by combining these singular patterns. Figure 4.11 outlines our approach, STAMP. In the first phase, the valid singular patterns are discovered. The second phase generates the candidates of valid complex pattern based on the candidates of valid singular patterns via triangle inequality. Finally, STAMP verifies all candidates, and finds the corresponding subsequence for each valid pattern so as to maximize its generalized information gain. The **maximum information gain (MIG)** counting is designed to determine whether an event a may participate in a pattern P of period l and can discover all valid singular patterns. However, the overall number of MIG counters could be quite large. As a result, it is beneficial if the number of MIG counters can be reduced to limit the number of scans through the event sequence. We, thus, propose a pruning technique, **optimal information surplus (OIS)**, to prune out disqualified periods of each event before the MIG counting. The OIS pruning and MIG counting constitute the first phase of STAMP. After MIG counting, the candidate complex patterns are generated, and then verified. We will explain each component in detail in the following sections.

MIG Counting

We first consider the issue of how to generate the MIG for a singular pattern, $(*, \ldots, a, *, \ldots, *)$, where the MIG serves as an estimate of the maximum achievable generalized information gain based on the maximum repetition of the singular pattern in a given sequence. We find that the problem of evaluating

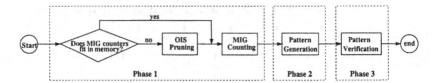

Figure 4.11. Outline of Stamp Approach

the MIG for a singular pattern on an event sequence $d_1, d_2, \ldots, d_N$, is equivalent to discover the maximum summation of any subsequence of a sequence of real numbers derived from the event sequence based on the singular pattern. The generalized information gain of an event sequence D with respect to a pattern P is $(S_D(P) - 1) \times I(P) - L_D(P)$. If P is a singular pattern $(*, \ldots, a, *, \ldots, *)$ of period l, then $I(P) = I(a)$. We can partition D into segments of length l. The information loss on a segment that does not support P is $I(a)$. Let m be the number of segments that do not support P. The generalized information gain of D with respect to P can be rewritten as $(S(P) - 1 - m) \times I(a)$. More specifically, each segment is associated with $I(a)$ if it supports P and $-I(a)$ otherwise. Therefore, we can map the event sequence D (with N events) into a sequence of $\lfloor \frac{N}{T} \rfloor$ real numbers $x_1, x_2, \ldots, x_{\lfloor \frac{N}{T} \rfloor}$. As shown in Figure 4.10, the sequence of real numbers with respect to $(a_4, *, *)$ is 1.33, -1.33, 1.33, 1.33, -1.33, 1.33, and 1.33. Now the problem becomes finding the maximum summation of any subsequence of $x_1, x_2, \ldots, x_{\lfloor \frac{N}{T} \rfloor}$. The maximum information gain problem can be then formulated as follows [4].

$$b(j + 1) = \max\{0, b(j)\} + x_{j+1}$$

$$c(j + 1) = \max\{b(j + 1), c(j)\}$$

where $b(j)$ and $c(j)$ are the maximum value of the summation of any sub-sequence ending exactly at x_j and the maximum value of the summation of any subsequence ending either before or at x_j, respectively. By choosing $b(0) = 0$ and $c(0) = 0$, the MIG of $(*, \ldots, a, *, \ldots, *)$ is equal to $\max\{c(\lfloor \frac{N}{T} \rfloor) - I(a), 0\}$. This forward recurrence can be easily solved by making one scan through the event sequence and maintaining two counters for each singular pattern to capture $b(.)$ and $c(.)$ at the current scanning position. The starting position and ending position of the corresponding subsequences can also be maintained simultaneously [4]. Since the number of MIG counters is $O(|\Im| \times L_{max}^2)$, and the main memory of a common computer is limited to a few hundred MBytes which could be far less than $O(|\Im| \times L_{max}^2)$, we need a mechanism to limit the number of MIG counters considered before the MIG counting procedure.

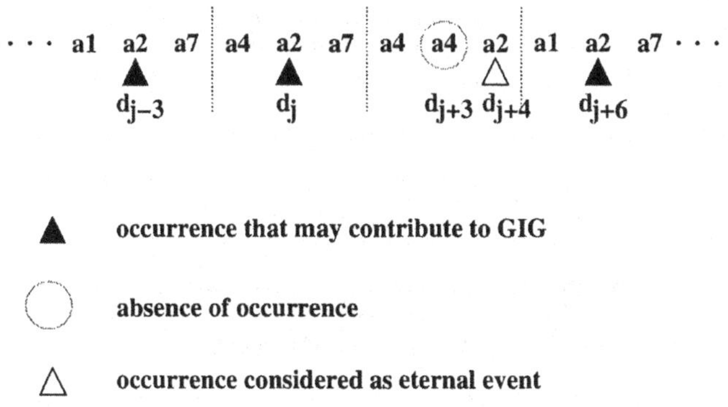

$$\cdots \text{ a1 } \text{ a2 } \text{ a7 } | \text{ a4 } \text{ a2 } \text{ a7 } | \text{ a4 } \textcircled{a4} \text{ a2 } | \text{ a1 } \text{ a2 } \text{ a7 } \cdots$$

▲				▲				△		▲
d_{j-3}				d_j				d_{j+3} d_{j+4}		d_{j+6}

▲   **occurrence that may contribute to GIG**

◯   **absence of occurrence**

△   **occurrence considered as eternal event**

Figure 4.12.   Occurrences of a_2

Segment-based OIS Pruning

To reduce the number of singular patterns for collecting the MIG counts, we introduce a pruning step OIS based on a concept to score consecutive periodic appearances of an event based on generalized information gain. OIS determines for each event, what are the likely periods (or equivalently for each period, what are the likely events that will have the period). For each likely period l of event a, there are l possible singular patterns, e.g., for $l = 3$, the three singular patterns are $(a, *, *)$, $(*, a, *)$, and $(*, *, a)$. We then use the MIG counters to screen these singular patterns. As we shall see the OIS storage requirement is $O(|\Im| \times L_{max})$ which is substantially lower than that of MIG counters and the OIS step can greatly reduce the candidates for MIG counters. Consider the occurrence of event a_k in an event sequence $D = d_1, d_2, \ldots$. Intuitively, a_k would generate the most information gain if it repeats itself perfectly at a certain period in a subsequence of D. If a_k participates in some pattern of period l, each occurrence of a_k could contribute to the generalized information gain of the pattern by an amount $I(a_k)$ in the optimistic scenario. However, in most case, the *net* information gain that a_k can contribute to a pattern is much smaller because (1) some information loss would incur due to the absence of a_k at some position; (2) some scattered/random occurrence of a_k has to be matched with the eternal event *. For example, the circled position in Figure 4.12 is considered as an absence of a_2 for pattern $(*, a_2, *)$. On the other hand, the third occurrences of a_2 have to be matched with the eternal event for pattern $(*, a_2, *)$. It is obvious that all of these suboptimal scenarios would weaken the net information gain that a_2 may contribute to any pattern. Since it is very complicated to track all of them, we choose to consider only one suboptimal scenario for estimating the information loss at this moment: the distance between two consecutive occurrences is greater than the length of the period. Let $Y(a_k, l)$ be the set of

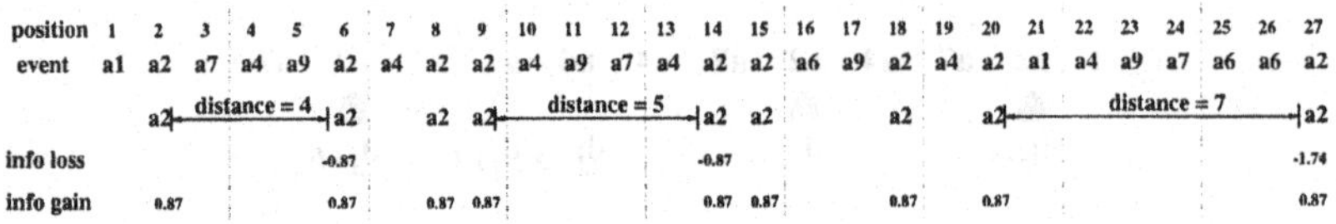

position	1	2	3	4	5	6	7	8	9	10	11	12	13	14	15	16	17	18	19	20	21	22	23	24	25	26	27
event	a1	a2	a7	a4	a9	a2	a4	a2	a2	a4	a9	a7	a4	a2	a2	a6	a9	a2	a4	a2	a1	a4	a9	a7	a6	a6	a2

a_2 ⊢ distance = 4 ⊣ a_2 a_2 a_2 ⊢ distance = 5 ⊣ a_2 a_2 a_2 a_2 ⊢ distance = 7 ⊣ a_2

info loss							-0.87							-0.87													-1.74
info gain	0.87					0.87			0.87	0.87					0.87	0.87			0.87		0.87						0.87

(a)

position	2	6	8	9	14	15	18	20	27
j	1	2	3	4	5	6	7	8	9
x(a2, 3, j)		-0.87			-0.87				-1.74
y(a2, 3, j)	0.87	0.87	0.87	0.87	0.87	0.87	0.87	0.87	0.87
f(a2, 3, j)	0.87	0.87	1.74	2.61	2.61	3.48	4.35	5.22	4.35
OIS(a2, 3, j)	0	0	0.87	1.74	1.74	2.61	3.48	4.35	3.48

(b)

Figure 4.13.   Optimal Information Surplus of Event a_2 for Period $l = 3$

patterns of period l which contains a_k. We employ the following disciplines to estimate the OIS that an event a_k might contribute to any pattern in $Y(a_k, l)$.

1 Each occurrence of a_k is considered to be a positive contribution by amount $I(a_k)$.

2 When the distance between any two consecutive occurrences is greater than l, information loss must incur due to the absence of a_k. In fact, there are at least $\lceil \frac{distance-l}{l} \rceil$ absences, one for each segment following the previous occurrence of a_k. Each absence would cause an information loss of $I(a_k)$.

Figure 4.13 shows the process to estimate the OIS that event a_2 could contribute to any pattern of period 3. Each occurrence introduces a generalized information gain. There are three places where the distances between two consecutive occurrences are 4, 5, and 7, respectively. Information loss of $I(a_2)$ incurs for each period in these subsequence(s) deficient in a_2.

These information losses and gains are essentially two lists of real numbers, namely $x(a_k, l, j)$ and $y(a_k, l, j)$ in Figure 4.13(b). At the jth occurrence of a_k, we can easily compute the the optimal net information surplus a_k could contribute in any event subsequence ending at that position, denoted by $OIS(a_k, l, j)$. Let $f(a_k, l, j)$ be the maximum net information aggregation of any subsequences ending exactly at the jth occurrence of a_k. We have

$$f(a_k, l, 0) = 0$$

$$f(a_k, l, j+1) = \max\{0, f(a_k, l, j) + x(a_k, l, j+1)\} + y(a_k, l, j+1)$$

$$OIS(a_k, l, j+1) = \max\{0, f(a_k, l, j+1) - I(a_k)\}$$

In the formula for $f(a_k, l, j+1)$, $\max\{0, f(a_k, l, j) + x(a_k, l, j+1)\}$ represents the contribution from subsequence ending at j and $y(a_k, l, j+1)$ represents the contribution from $(j + 1)$th position. Note that since $x(a_k, l, j + 1)$ captures the potential information loss in the portion of subsequence prior to $(j + 1)$th position and should not affect the subsequence starting at $(j + 1)$th position, $x(a_k, l, j + 1)$ and $y(a_k, l, j + 1)$ are treated differently in the above formula. A linear algorithm would compute all OIS values as illustrated in Figure 4.13(b). Note that the OIS is an optimistic estimation and only gives an upperbound of the generalized information gain that an event a_k would contribute to any pattern. It is obvious that event a_2 does not exhibit strong pattern in Figure 4.13(a). However, the above OIS pruning method overestimates the generalized information gain of a_2 in Figure 4.13(b). Therefore, we propose another more sophisticated, but more effective OIS pruning method at the end of this section.

MIG Counter Generation after OIS Pruning

After obtaining OIS values, for each period length l, we want to find which event is likely to appear in a valid pattern of period l. Let E_l denote the set of such events. By definition, any pattern of period l may contain at most l different events. The problem can be converted to testing whether the combined OIS of a set of l events may exceed the generalized information gain threshold g at some position in the sequence. Even though there are totally $\binom{|\Im|}{l}$ different event combinations, it is not necessary to examine all of them. Conceptually, each event in a valid pattern must play a supporting role in accumulating generalized information gain. Therefore, we only need to consider the set of events with positive OIS at any time. (Note that this set may vary over time.) As we mentioned before, the event sequence can be treated as a list of segments of length l as shown in Figure 4.14. Each segment might serve as the last segment of a valid subsequence for some pattern. E_l need to be updated at the end of each segment. Let $T(l, s)$ be the set of events with positive OIS at the end of the sth segment, i.e., $T(l, s) = \{a_k \mid OIS(a_k, l, j_k) > 0, j_k$ is the last occurrence of a_k before the end of the sth segment $\}$. For example, a_2 and a_4 are the only events with positive OIS value in segment 3, i.e., $T(3, 3) = \{a_2, a_4\}$. Since the OIS value of an event a_k is updated for each occurrence of a_k, it might not be updated in every segment and might also be updated multiple times within a segment. In any case, we always use the most recent value for the computation. In Figure 4.14, the OIS value of a_2 is not updated in segment 4 and is updated twice in segment 5. Then the OIS value that we use for these two segments are 1.74 and 2.61, respectively.

For each segment s, let $E(l, s)$ be the set of events that may appear in a pattern whose valid subsequence ends at the sth segment. $E(l, s)$ is essentially a subset of $T(l, s)$ and can be computed easily[12]. After we calculate $E(l, s)$ for all segments, the set E_l can be trivially obtained by taking the union, i.e., $E_l = \bigcup_{\forall s} E(l, s)$. Figure 4.14 shows the process to compute the candidate

	a1 a2 a7	a4 a9 a2	a4 a2 a2	a4 a9 a7	a4 a2 a2	a6 a9 a2	a4 a2 a1	a4 a9 a7	a6 a6 a2
s	1	2	3	4	5	6	7	8	9
OIS a1	0							0	
a2	0	0	0.87 1.74		1.74 2.61	3.48	4.35		3.48
a4		0	1.33	2.66	3.99		3.99	5.32	
a6						0			0 1.03
a7	0			0				0	
a9		0		0		0		0	
T(3, s)	{}	{}	{a2, a4}	{a2, a4}	{a2, a4}	{a2, a4}	{a2, a4}	{a2, a4}	{a2, a6}
E(3,s)	{}	{}	{}	{a2, a4}	{a2, a4}	{a2, a4}	{a2, a4}	{a2, a4}	{a2, a6}
E_3	{}	{}	{}	{a2, a4}	{a2, a4}	{a2, a4}	{a2, a4}	{a2, a4}	{a2, a4, a6}

Figure 4.14. Candidate Event(s) Generation for Period $l = 3$ and Minimum Generalized Information Gain $g = 3.5$

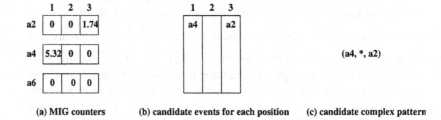

(a) MIG counters   (b) candidate events for each position   (c) candidate complex pattern

Figure 4.15. MIG Counters and Candidate Patterns for Period $l = 3$ and Minimum Generalized Information Gain $g = 3.5$

events for period l. Note that a single scan of the event sequence is sufficient to compute the candidate events for all possible periods. For any given period l, if the candidate event set E_l is empty, then no pattern of period l would be able to meet the required generalized information gain. Once E_l is determined, for each event in E_l, we proceed to determine the singular pattern candidates with period l using the MIG counters. A counter is initialized for each event $a_k \in E_l$ at each position in the period. There are totally $l \times \mid E_l \mid$ counters where $\mid E_l \mid$ is the cardinality of E_l. For example, $E_3 = \{a_2, a_4, a_6\}$ for period 3. There are 3 different positions an event may occupy in a period. In turn, 3 MIG counters are initialized for each event, one for each position as illustrated in Figure 4.15(a). The procedure presented in the previous subsection is then carried out to collect all these MIG counters.

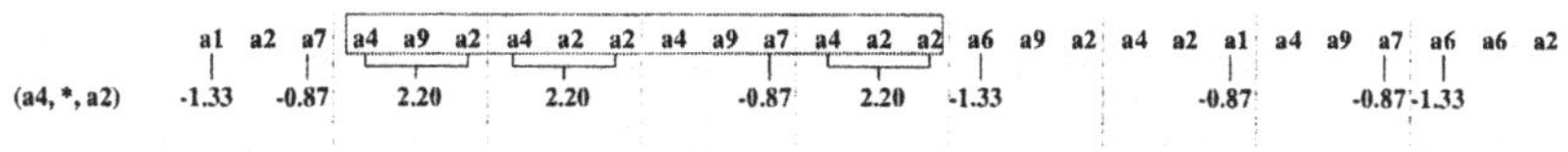

a1 a2 a7 | a4 a9 a2 | a4 a2 a2 | a4 a9 a7 | a4 a2 a2 | a6 a9 a2 | a4 a2 a1 | a4 a9 a7 | a6 a6 a2

(a4, *, a2) -1.33 -0.87 2.20 2.20 -0.87 2.20 -1.33 -0.87 -0.87 -1.33

Figure 4.16. Verification of pattern $(a_4, *, a_2)$

Candidate Pattern Generation

After all these MIG counters have been gathered, for each position of a period (of length l), we keep all events with positive MIG as the candidate events for this position as shown in Figure 4.15(b). The candidate complex patterns[13] of period l can be generated iteratively. A pattern $(p_1, p_2, \ldots, p_l)$ is constructed each time by assigning each position p_j either an event in the corresponding candidate event set or eternal event. Let $MIG(p_j)$ is the MIG count for the singular pattern $(*, \ldots, p_j, *, \ldots)$ with p_j at the jth position. If $\sum_{1 \leq j \leq l} MIG(p_j) \geq g$, this pattern will be taken as a candidate pattern to the verification process presented later in this section. Otherwise, this generated pattern is simply discarded since it is impossible for this pattern to meet the generalized information gain threshold g. For example, Figure 4.15(c) shows the set of candidate patterns generated from the candidate events in Figure 4.15(b) with the threshold $g = 3.5$.

Pattern Verification

The verification process of a candidate pattern $P = (p_1, p_2, \ldots, p_l)$ of period l is similar to that to compute the MIG counts. The event sequence $d_1, d_2, \ldots$ is first mapped to a list of real numbers as follows. Each segment of l events $d_{l \times j+1}, d_{l \times j+2}, \ldots, d_{l \times j+l}$ is examined at a time. It is mapped to a positive number $I(P)$ if it supports P. Otherwise, a negative number $-I(p_k)$ is mapped from each violated position of p_k. Then, a similar procedure to MIG computing can be applied to locate the subsequence that maximize the generalized information gain of the pattern. Figure 4.16 shows the validation of pattern $(a_4, *, a_2)$. The bracketed subsequence is the one that provides the maximum generalized information gain 3.53. Note that multiple subsequences may have the same generalized information gain. If that is the case, our algorithm will output the first one[14].

Sophisticated OIS Pruning with Superfluous Gain Elimination

In this section, we discuss some techniques that can provide a better estimation of the value OIS. In section 3.2, we only consider the case that the gap between two consecutive occurrences of an event a_k exceeds the period when the information loss is calculated. We now examine the case that the gap is less than the period. Let's first consider the scenario that a_k repeats itself perfectly at certain distance l in a subsequence D'. For every occurrence of a_k (except

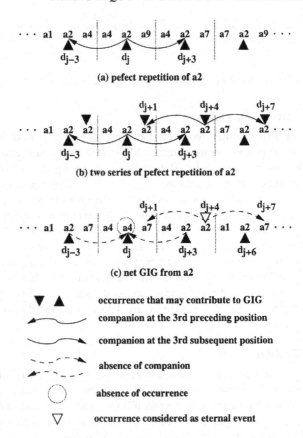

Figure 4.17. (Revisit) Occurrences of a_2

the first and last one) at position j within D', we can observe two companion occurrences of a_k at position $j - l$ and $j + l$, referred to as the **backward** and **forward** companion, respectively. Figure 4.17(a) shows a series of 4 occurrences of event a_2 with period 3, each of which (e.g., at position d_j) has two companions at the third preceding position (e.g., d_{j-3}) and the third subsequent position (e.g., d_{j+3}) of the current one. The total generalized information gain of the pattern $(*, a_2, *)$ generated from the subsequence is $(4 - 1) \times I(a_2)$. The above claim also holds in the case where an event occurs multiple times within a period. Figure 4.17(b), shows two overlapping series of repetitions of a_2 which may bring $(4 - 1) \times 2 \times I(a_2)$ generalized information gain to pattern $(*, a_2, a_2)$.

As we mentioned before, the *net* information gain a_k can contribute to a pattern is much confined if one of the following is true.

(1) Some information loss would incur due to the absence of a_k at some position. This absence can be viewed as a companion absence of the neighboring occurrence(s). For example, the circled position in Figure 4.17(c) is considered as an absence of a_2 for pattern $(*, a_2, *)$. This absence makes both position d_{j-3} and position d_{j+3} lose one companion.

(2) Some misaligned occurrence of a_k has to be treated as an eternal event or even as an absence of some other event and hence no generalized information gain can be collected from it. These occurrences usually are associated with companion absences. The third occurrences of a_2 in Figure 4.17(c) have to be treated as an eternal event for pattern $(*, a_2, *)$ and treated as "absence of a_7" for pattern $(*, a_2, a_7)$. We also observe that in Figure 4.17(c) the event a_2 at position d_{j+4} has no companion at both position d_{j+1} and position d_{j+7}.

It is obvious that all of these suboptimal scenarios would weaken the net generalized information gain that a_2 may contribute to any pattern. By carefully identifying *companion absence* and assessing its impact to the generalized information gain, we can obtain a tighter estimation of the net contribution to the generalized information gain that a_k could make to any pattern in $Y(a_k, l)$. Since the companionship is essentially a mutual relationship, we choose to examine the backward companion to avoid looking forward in the event sequence. This is achieved by employing an additional discipline to estimate the OIS that an event a_k might contribute to any pattern in $Y(a_k, l)$.

(3) For each occurrence of a_k at position d_j, if the distance to the previous occurrence of a_k is less than l, then the previous occurrence of a_k is not the backward companion of a_k at position d_j. So an information adjustment of $-I(a_k)$ is needed.

The new calculation of OIS is as follows.

$$
\begin{aligned}
f(a_k, l, 0) &= 0 \\
f(a_k, l, j+1) &= \max\{e(a_k, l, j), \\
&\quad f(a_k, l, j) + x(a_k, l, j+1) \\
&\quad + y(a_k, l, j+1) + z(a_k, l, j+1)\} \\
OIS(a_k, l, j+1) &= f(a_k, l, j+1) - I(a_k)
\end{aligned}
$$

where $e(a_k, l, j)$ is the product of $I(a_k)$ and the number of occurrences of a_k in the previous l events prior to the jth occurrence of a_k. Here $z(a_k, l, j)$ is the adjustment according to the third discipline stated above. The revised computation of OIS is shown in Figure 4.18(a). For each period length l, the addition storage requirement to perform the generalized information gain adjustment is an array of l elements (to store the previous l events). The computation complexity remains the same.

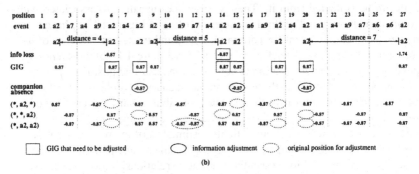

position	2	6	8	9	14	15	18	20	27
j	1	2	3	4	5	6	7	8	9
$x(a2,3,j)$		-0.87			-0.87				-1.74
$y(a2,3,j)$	0.87	0.87	0.87	0.87	0.87	0.87	0.87	0.87	0.87
$z(a2,3,j)$			-0.87			-0.87		-0.87	
$e(a2,3,j)$	0.87	0.87	1.74	1.74	0.87	1.74	0.87	1.74	0.87
$f(a2,3,j)$	0.87	0.87	1.74	2.61	2.61	2.61	3.48	3.48	2.61
$OIS(a2,3,j)$	0	0	0.87	1.74	1.74	1.74	2.61	2.61	1.74

(a)

(b)

Figure 4.18. Sophisticated OIS Estimation of Event a_2 with respect to Period $l = 3$

Finally, we give some rationale for the third discipline. Intuitively, when the companion absence is present, the generalized information gain would not be as high as expected. Some adjustment needs to be taken to provide a tighter estimation. Let's reconsider the example shown in Figure 4.13(a), which is also described in Figure 4.18(b). $(*, a_2, *)$, $(*, *, a_2)$, and $(*, a_2, a_2)$ are the three possible patterns that involve a_2 because a_2 only appears at the second and third position on each segment.

- The adjustment in position 8 comes from the following reasons. The solid ovals indicate the adjustments that are taken according to the third discipline. For $(*, a_2, *)$, comparing the actual information gain/loss with the previous estimation in section 3.2, the generalized information gain on position 6 is superfluous. For $(*, *, a_2)$, the generalized information gain on position 8 is superfluous, while the information on position 6 for pattern $(*, a_2, a_2)$ is superfluous. Therefore, by position 8, one generalized information gain of 0.87 is superfluous for all patterns, thus, we adjust the generalized information gain by -0.87.

- The adjustment in position 15 is necessary because of the following reasons. Generalized information gain on position 14 and 15 is superfluous for pattern $(*, *, a_2)$ and $(*, a_2, *)$, respectively, due to the similar reasons described above. For $(*, a_2, a_2)$, on position 11 and 12, we need to deduct generalized information gain by 1.74; however, we only deduct generalized information gain by 0.87 on position 14. As a result, an additional 0.87 needs to be deducted from the net generalized information gain. Thus, we add an adjustment of -0.87 on position 15.

- On position 20, another adjustment is created. There are two superfluous information gain of 0.87 on position 18 and 20 for $(*, a_2, a_2)$. Also, there is one superfluous information gain of 0.87 by position for both $(*, a_2, *)$ and $(*, *, a_2)$ as indicated in Figure 4.18(b). Therefore, one generalized information gain adjustment of -0.87 is created on position 20.

As a rule of thumb, for an event a_k, the adjustment is postponed until the time it would apply to all possible patterns involving a_k. Therefore, the new estimation could be used as a tighter bound on the generalized information gain.

Experimental Results

We implemented the STAMP in C programming language on an IBM RS-6000 (300 MHz CPU) with 128MB running AIX operating system. In the following experiments, we set $L_{max} = 1000$.

Synthetic Sequence Generation

For the purpose of evaluation of the performance of STAMP, we use four synthetically generated sequences. Each sequence consists of 1024 distinct events and 20M occurrences of events. The synthetic sequence is generated as follows. First, at the beginning of the sequence, the period length l of the next pattern is determined, which is geometrical distributed with mean μ_l. The number of events involved in a pattern is randomly chosen between 1 and l. The number of repetitions m of this pattern is geometrical distributed with mean μ_m. The events that are involved in the pattern are chosen according to a normal distribution with mean $\frac{1}{1024}$ (there are total 1024 distinct events) and standard deviation 2. However, the pattern may not perfectly repeat itself for m times. To simulate the imperfectness of the subsequence, we employ a parameter δ to control the noise. δ is uniformly distributed between 0.5 and 1. With probability δ, the next l events match the pattern. Otherwise, the next l events do not support the pattern. The replacement events are chosen from the event set with the same normal distribution (mean and standard deviation equal to $\frac{1}{1024}$ and 2, respectively). This subsequence ends when there are m matches, and a new subsequence for a new pattern starts. This process repeats until it reaches the end of the sequence. Four sequences are generated based on values of μ_l and μ_m in Table 4.5.

Data Set	μ_l	μ_m	Distinct events	Total Events
$l3m20$	3	20	1024	20M
$l100m20$	100	20	1024	20M
$l3m1000$	3	1000	1024	20M
$l100m1000$	100	1000	1024	20M

Table 4.5. Parameters of Synthetic Data Sets

Effects of OIS Pruning

Figure 4.19 (a) shows the difference of the pruning power of the sophisticated OIS pruning with superfluous information gain elimination and the segment-based OIS pruning. Since the behaviors are similar with different event sequences, we only show the pruning results for sequence $l3m20$. It is evident that the sophisticated one is much more effective. Although the more sophisticated OIS pruning requires a little bit more space and time, the result is improved dramatically. Therefore, we decide that in the remainder of this section, we use the implementation of the more sophisticated OIS pruning technique in STAMP.

Figure 4.19 (b) shows the effectiveness of the more sophisticated OIS pruning. The y-axis shows the fraction of the MIG counters that would not be needed. It is evident that when the generalized information gain threshold g increases, the OIS is more effective because less events at each period may qualify for MIG counting. However, even with $g = 2$, the OIS pruning can filter out more than 50% of the MIG counters.

Effects of MIG Counting

The number of candidate patterns depends on the average number of events (with positive MIG values) in each position of each period. Figure 4.20(a) shows the average number of events (α) in each position for period (l) between 5 and 100 with the generalized information gain threshold $g = 5$. (Note that the Y-axis is in log scale.) The α value is similar for all four data sets and α decreases with l. In Figure 4.20(a), when $l > 30$, $\alpha < 1$ for all four sequences. In other words, many positions of a pattern with period larger than 30 are null. The total number of candidate patterns (β) for each period between 5 and 100 is illustrated in Figure 4.20(b). β increases with l when $l < 30$ due to the longer periods. On the other hand, β decreases with l when $l > 30$ due to the smaller number of possible events in each position.

Overall Performance

The overall performance of STAMP depends largely on the number of MIG counters and candidate patterns. If MIG counters and candidate patterns can not fit into main memory, then multiple scans of the event sequence is needed to

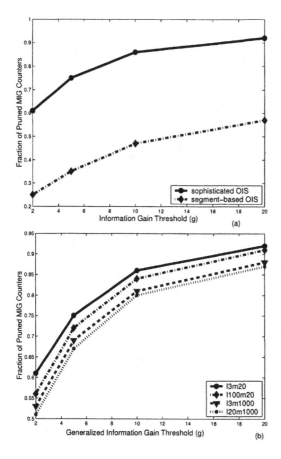

Figure 4.19. Effects of OIS Pruning

generate the counters or verify the candidate patterns. For all four sequences, after the OIS counting, the MIG counters can fit into the main memory. However, the candidate patterns can not fit into main memory at once. One third of the candidate patterns of $l100m1000$ sequence can be loaded into memory each time while half of the candidate patterns of the other three sequences can be loaded into memory each time for $g = 5$. Figure 4.21(a) shows the overall response time of STAMP for four event sequences with respect to the generalized information gain threshold. The average performance of $l2m20$, $l100m20$, and $l3m100$ is similar because of the similar number of disk I/Os whereas the performance of $l100m1000$ is significantly higher.

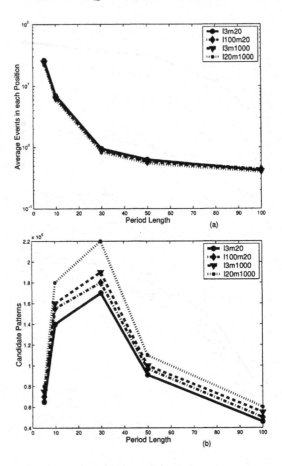

Figure 4.20. Candidate Patterns

To analyze the benefits of using the generalized information gain as a threshold versus using support and confidence as thresholds, we also implemented a data mining tool which finds the subsequences with maximum support while maintaining the confidence and support above certain thresholds. We call this subsequence discovery tool *s-c Miner*. In s-c Miner, an approach similar to [12] is used to generate all patterns that satisfy the support threshold followed by an algorithm adapted from [10] to find the desired subsequence. Figure 4.21 (b) shows the performance difference between STAMP and s-c Miner. (Note the y-axis in Figure 4.21 (b) is in log scale.) Since the performance on all four sequences is similar, thus, we only show the performance of sequence $l3m20$. The support and confidence thresholds in s-c Miner is set in such a way that all

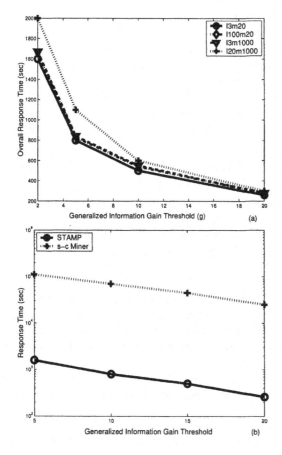

Figure 4.21. Performance of STAMP

subsequences found by STAMP can also be found by s-c Miner. Since a large number of patterns are generated by s-c Miner which are not deemed as valid by STAMP, the performance of s-c Miner is severely impact. However, readers should keep in mind that in some applications, if the support and confidence are the proper measurement to qualify a valid pattern, then the s-c Miner should be preferred.

Notes

1 For a pattern s in a sequence $d_1, d_2, \ldots, d_N$, the frequency count is defined as $| \{i \mid 0 \leq i \leq \frac{N}{|s|}, \text{and the string } s \text{ is true in } d_{i|s|+1}, \ldots, d_{i|s|+|s|}\} |$.

2 This requirement is employed to exclude the trivial pattern $(*, *, \ldots, *)$ from being considered.

3 Given two sequences D and D', D is a **subsequence** of D' if D can be obtained by removing some events in D'.

4 A candidate pattern is a pattern that can be constructed by assigning each open position an event from the corresponding candidate list of that position.

5 The adjusted accuracy is often used when the class distribution is very skewed. Given two classes $normal$ and $abnormal$, the adjusted accuracy is defined as $\dfrac{\frac{Pred(normal) \cap Act(normal)}{Act(normal)} + \frac{Pred(abnormal) \cap Act(abnormal)}{Act(abnormal)}}{2}$ where $Pred(X)$ is the percentage of time the classifier predict X and $Act(X)$ is the percentage of time the class is X.

6 For example, the GIG of $(a_1, a_2, *)$ can not exceed the summation of that of $(a_1, *, *)$ and $(*, a_2, *)$.

7 For the sake of simplicity of exploration, we assume that, without additional qualification, a_j occurs equally likely at any position with probability $Prob(a_j)$. All results presented in this chapter can be modified to apply to a more general scenario.

8 Another way of looking at it is that $Prob(*) = 1$ at any time.

9 Here we choose the number of distinct events in the sequence as the base for calculating the information. It is inconsequential what is the base as long as the generalized information gain threshold specified by the user is consistent with the base.

10 Since we aim at mining periodic patterns, only repeated occurrences of a pattern are used to accumulate the generalized information gain.

11 We will use the information shown in Figure 4.9 constantly in all subsequent examples in this chapter.

12 One way to compute it is to examine the events in $T(l, s)$ in descending order of their OIS values. $E(l, s)$ is $\emptyset$ if the combined OIS of the l largest ones is below g. Otherwise, all of these l events (with largest OIS values) are put into $E(l, s)$; and each remaining event $a_k \in T(l, s)$ is also added to $E(l, s)$ until the combined OIS of a_k and the $(l - 1)$ largest ones is below g.

13 All singular patterns have already been considered in the MIG counting procedure.

14 With minor modification of the algorithm, all such subsequences can be output.

References

[1] Agrawal, R., and Srikant, R. (1994). Fast algorithms for mining association rules in large databases. *Proc. of the Int'l Conference on Very Larg Databases (VLDB)*. pp. 487-499.

[2] Agrawal, R., and Srikant, R. (1995). Mininig sequential patterns. *Proc. of the Int'l Conference on Data Engineering (ICDE)*. pp. 3-14.

[3] Berger, G., and Tuzhilin, A. (1998). Discovering unexpected patterns in temporal data using temporal logic. *Temporal Databases - Research and Practice, Lecture Notes on Computer Sciences*. vol. (1399) pp. 281-309.

[4] Bentley, J. (1984). Programming pearls. *Communications of ACM*. (27)2:865-871.

[5] Blahut, R. (1987). *Principles and Practice of Information Theory*. Addison-Wesley Publishing Company.

[6] Brin. S., Motwani, R., and Silverstein, C. (1997). Beyond market baskets: generalizing association rules to correlations. *Proc. ACM SIGMOD Int'l. Conference on Management of Data (SIGMOD)*. pp. 265-276.

[7] Califano, A., Stolovitzky, G., and Tu, Y. (1999) Analysis of gene expression microarrays: a combinatorial multivariate approach, *IBM T. J. Watson Research Report*.

[8] Chakrabarti, S., Sarawagi, S., and Dom, B. (1998). Mining surprising patterns using temporal description length. *Proc. Int. Conf. on Very Large Data Bases (VLDB)*. pp. 606-617.

[9] Cohen, E., Datar, M., Fuijiwara, S., Cionis, A., Indyk, P., Motwani, R., Ullman, J., and Yang, C. (2000). Finding interesting associations without support pruning. *Proc. 16th IEEE Int'l. Conference on Data Engineering (ICDE)*. pp. 489-499.

[10] Fukuda, T., Morimoto, Y., Morishita, S., and Tokuyama, T. (1996). Mining optimized association rules for numeric attributes. *Proc. 15th ACM Symposium on Principles of Database Systems (PODS)*. pp. 182-191.

[11] Han, J., Gong, W., and Yin, Y. (1998). Mining segment-wise periodic patterns in time-related databases. *Proc. Int'l. Conference on Knowledge Discovery and Data Mining (KDD)*. pp. 214-218.

[12] Han, J., Dong, G., and Yin, Y. (1999). Efficient mining partial periodic patterns in time series database. *Proc. IEEE Int'l. Conference on Data Engineering (ICDE)*. pp. 106-115.

[13] Liu, B., Hsu, W., and Ma, Y. (1999) Mining association Rules with multiple minimum supports. *Proc. Int'l. Conference on Knowledge Discovery and Data Mining (KDD)*. pp. 337-341.

[14] Mannila, H., Pavlov, D., and Smyth, P. (1999). Prediction with local patterns using cross-entropy. *Proc. Int'l. Conference on Knowledge Discovery and Data Mining (KDD)*. pp. 357-361.

[15] Oates, T., Schmill, M., and Cohen, P. (1999). Efficient mining of statistical dependencies. *Proc. 16th Int. Joint Conf. on Artificial Intelligence*. pp. 794-799.

[16] Ozden, B., Ramaswamy, S., and Silberschatz, A. (1998). Cyclic association rules. *Proc. 14th Int'l. Conference on Data Engineering (ICDE)*. pp. 412-421.

[17] Silberschatz, A., and Tuzhilin, A. (1996). What makes patterns interesting in knowledge discover systems. *IEEE Transactions on Knowledge and Data Engineering (TKDE)*. vol. 8 no. 6, pp. 970-974.

[18] Yang, J. Wang, W., and Yu, P. (2001). Infominer: mining surprising periodic patterns. *Proc. of the Seventh ACM Int'l Conference on Knowledge Discover and Data Mining (KDD).* pp. 395-400.

[19] Yang, J. Wang, W., and Yu, P. (2003). STAMP: on discovery of statistically important pattern repeats in long sequential data. *Proc. of the Third SIAM International Conference on Data Mining (SDM).*

[20] Yang, J., Wang, W., and Yu, P. (2003). Mining asynchronous periodic patterns in time series data. *IEEE Transactions on Knowledge and Data Engineering.* 15(3):613-628.

[21] Zaki, M. (2000). Generating non-redundant association rules. *Proc. of the Sixth ACM Int'l Conference on Knowledge Discover and Data Mining (KDD).* pp. 34-43.

Chapter 5

APPROXIMATE PATTERNS

As mentioned before, the data is often of poor qualities. In other words, the symbols in the input data may be inaccurate. There are two sources of noises or inaccuracy in the data. The first one is the error or imperfection in the data collection process. For instance, in assembling a DNA sequence, there exists uncertainty of which base pair should assigned in each position. Therefore, a score is associated with each base pair (symbol) to indicate the confidence of the symbol. The second type of noises is the system noise, which is inherent in the underlying system. The mutation of the amino acids in a protein sequence is one of such examples. In nature, an amino acid may mutate to another without changing the biological function of the protein. In this chapter, we will describe two approaches that deal with the noises in the input data.

1.   Obscure Patterns

Due to the presence of noise, a symbol may be misrepresented by some other symbols. This substitution may prevent an occurrence of a pattern from being recognized and in turn slashes the support of that pattern. As a result, a frequent pattern may be "concealed" by the noise. This phenomenon commonly exists in many applications.

- *Bio-Medical Study*. Mutation of amino acids is a common phenomenon studied in the context of biology. Some mutations are proved to occur with a non-negligible probability under normal circumstances and incur little change to its biological functionalities. For example, the amino acid N is likely to mutate to D with little impact to the behavior [7]. In this sense, they should not be considered as totally independent individuals.

- *Consumer Behavior*. It happens frequently that a customer may end up buying a slightly different merchant from what he (she) originally wanted

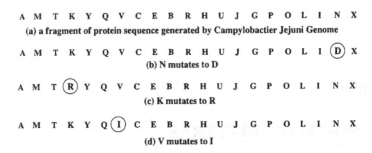

Figure 5.1. An Example of Protein Sequence

due to various reasons, such as the desired one was out of stock or misplaced. Allowing obscurity in item matching may conduce to unveil the customer's real purchase intention.

This problem becomes critical when the pattern is long because a long pattern is much more vulnerable to distortion caused by noise. Our experiments show that, even with a moderate degree of noise, a frequent long pattern may have as much as 60% chance to be labeled as an infrequent pattern. In addition, the set of patterns that fail to be detected using the support model may be very crucial. For example, our experiments also show that the loss of these vital patterns may significantly degrade the performance of the classifier built upon the set of frequent patterns.

Let's take the gene sequence analysis as an example. The length of a gene expression can range up to a few thousands if amino acids are taken as the granularity of the analysis. Figure 5.1(a) shows a fragment of a protein sequence that is found in campylobacter jejuni genome [20]. Some clinical studies show that, the amino acids N, K, and V are relatively more likely to mutate to amino acids D, R, and I, respectively. The corresponding gene expressions after the mutation are shown in Figure 5.1(b), (c), and (d) respectively. Even though all of these mutated gene expressions somewhat differ from the standard one in Figure 5.1(a), it is more equitable to treat them as possible (degraded) occurrences of the standard expression than to consider them as totally independent gene expressions.

In order to accommodate the above circumstance, it is necessary to allow some flexibility in pattern matching. Unfortunately, most previously proposed models (as discussed before) for sequential patterns only take into account exact match of the pattern in data. In this section, we present a more flexible model that allows obscurity in pattern matching [28]. A so-called *compatibility matrix* is introduced to enable a clear representation of the likelihood of symbol substitutions. Each entry in the matrix corresponds to a pair of symbols (x, y) and specifies the conditional probability that x is the true value given

observed value / true value	d1	d2	d3	d4	d5
d1	0.9	0.1	0	0	0
d2	0.05	0.8	0.05	0.1	0
d3	0.05	0	0.7	0.15	0.1
d4	0	0.1	0.1	0.75	0.05
d5	0	0	0.15	0	0.85

Figure 5.2. An Example of Compatibility Matrix

y is observed. Figure 5.2 gives an example of the compatibility matrix. The compatibility matrix essentially creates a natural bridge between the observation and the underlying substance. Each observed symbol is then interpreted as an occurrence of a set of symbols with various probabilities. For example, an observed d_1 corresponds to a true occurrence of d_1, d_2, and d_3 with probability 0.9, 0.05, and 0.05, respectively. Similarly, an observed symbol combination is treated as an occurrence of a set of patterns with various degrees. A new metric, namely *match*, is then proposed to quantify the significance of a pattern and is defined as the "aggregated amount of occurrences" of a pattern in the sequence database. The match of a pattern indeed represents the "real support" that was expected if no noise presents.

The well-known Apriori property also holds on the match measure, which states that *any superpattern[1] of an infrequent pattern is also infrequent and any subpattern of a frequent pattern is also frequent.* This guarantees that any previous algorithm designed for the support model [1, 10, 13, 18] can be generalized to suit the match model, though it may not necessarily be efficient. Compared to the support model, a much larger number of patterns may possess some positive matches. In addition, the length of a pattern can be considerably long in the context we address, e.g., gene expression analysis. The combined effect of these two factors may force any direct generalization of existing algorithms (even including those designed for long patterns [4]) to scan the entire sequence database many times. To tackle this problem, we propose a novel sampling-based algorithm that utilizes the Chernoff Bound [6, 14, 23, 25] to estimate the set of patterns whose matches in the sample are very close to the threshold so that there is no sufficient statistical confidence to tell whether the pattern would be frequent or not in the entire sequence database. These ambiguous patterns are then investigated against the entire sequence database to finalize the set of frequent patterns. Because the sample size is usually limited by the memory capacity and the distribution-independent nature of Chernoff bound provides a very conservative estimation, the number of ambiguous patterns is usually very large. Consequently, significant amount of computation needs to be consumed in order to verify these ambiguous patterns in a level-wise manner. We observed that, for the protein sequence database, majority of the time would be spent in this verification step when the discovered pattern contains dozens of symbols. To expedite the process, we proposed a so called *border collapsing* technique to conduct the examination of these ambiguous patterns. While the super-pattern/sub-pattern relationship forms a lattice among all patterns, the set of ambiguous patterns "occupies" a contiguous portion of the lattice according to the Apriori property. Therefore, starting from the lower border and the upper border embracing these ambiguous patterns, the border of frequent patterns (in the entire sequence database) is located efficiently by successively collapsing the gap between these two borders until no ambiguous pattern exists. To maximize the extent of each gap collapsing operation, only the set of ambiguous patterns with the highest collapsing power are identified and probed. As a result, the expected number of scans through the entire database is minimized.

There is a clear distinction between our algorithm and existing algorithms [23, 25] that also use sampling technique to mine frequent patterns. In the previous proposed approaches, the frequent patterns calculated from the sample is usually taken as the starting position of a level-wise search conducted in the entire sequence database until all frequent patterns have been identified. This strategy is efficient if the number of frequent patterns that fail to be recognized from the sample is small, which is typically the case under the assumption of a reasonably large sample size and a relatively short pattern length. However, in the problem we try to solve, the number of ambiguous patterns may be substantially larger, which makes a level-wise search an inefficient process. In contrast, our algorithm can successfully deal with such scenario by each time directly probing the set of ambiguous patterns that would lead to a collapse of the space of remaining ambiguous patterns to the largest extent, so that the number of necessary scans through the sequence database is minimized. This leads to substantially better performance than the existing sampling approach. We will investigate the effect of the border collapsing technique in more detail in a later section and will show that, in most cases, several scans of the sequence database are sufficient even the pattern is very long when the border collapsing is employed.

Model of Obscure Patterns

Here we are interested in finding patterns that may be concealed (to some extent) by noise in a sequence database. We first introduce some terminologies that will be used throughout this chapter. Let Θ be a set of distinct symbols $\{d_1, d_2, \ldots, d_m\}$. **Definition** A **sequence** of **length** l is an ordered list of l symbols in Θ. A **sequence database** is a set of tuple $\langle Sid, S \rangle$ where Sid is the ID of the sequence S.

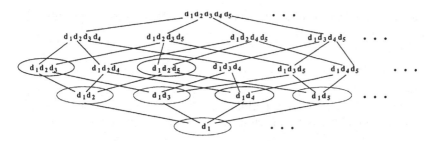

Figure 5.3. A Fragment of Lattice of Sequential Patterns

Definition A **pattern** of **length** l is represented as a list of l symbols, each of which is a symbol in Θ. A pattern of length l is also referred to as a l-**pattern**.

Note that there is no formative difference between a sequence and a pattern. Conventionally, we use *sequence* to refer to the raw data in the database which serves as the input to the data mining algorithm and use *pattern* to denote *subsequence* appearing in the database (which may become the output produced by the algorithm). Note that both the length of a sequence and the length of a pattern may be very large even though the alphabet Θ is very limited. For example, a protein sequence typically contains hundreds (if not thousands) of amino acids from an alphabet of 20 different amino acids.

Given a sequence $S = S_1 S_2 \ldots S_{l_S}$, a pattern $P = d_1 d_2 \ldots d_{l_P}$ is a **sub-sequence** of S if there exist a list of integers $1 \leq i_1 \leq i_2 \leq \cdots \leq i_{l_P} \leq l_S$ such that $d_j = S_{i_j}$ for $1 \leq j \leq l_P$. Given two patterns $P = d_1 d_2 \ldots d_l$ and $P' = d'_1 d'_2 \ldots d'_{l'}$ where $l' \leq l$, P' is a **subpattern** of P if there exists a list of integers $1 \leq i_1 \leq i_2 \leq \cdots \leq i_{l'} \leq l$ such that $d'_j = d_{i_j}$ for $1 \leq j \leq l'$. In such a case, P is also referred to as a **super-pattern** of P'. Intuitively, P' is a subpattern of P if P' can be generated by dropping some portion(s) of P^2. For example, $d_1 d_3$ and $d_1 d_4 d_5$ are subpatterns of $d_1 d_3 d_4 d_5$ but $d_1 d_2$ is not. It is clear that the sub-/super-pattern relationship defines a lattice among all patterns. Figure 5.3 shows a fragment of the lattice.

Our goal is to find the significant patterns in a sequence database in the presence of noise. In order to accommodate the noise, we propose a flexible model that allows obscurity in pattern matching. If the observed data does not match exactly but is somewhat "compatible" with a pattern, it can be regarded as a degraded occurrence of the pattern. To honor the "partial" occurrence of a pattern, we propose a new metric, namely *match*, to characterize the significance of the pattern in a symbol sequence. In particular, the conditional probability of the true value given an observed symbol is utilized to quantify "compatibility" between a pattern and an observed symbol sequence, and to assess the *match* of the pattern.

Definition Let $\Theta = \{d_1, d_2, \ldots, d_m\}$ be a set of distinct symbols. An $m \times m$ matrix C, referred to as **compatibility matrix**, can be used to represent the conditional probabilities for each pair of symbols. Given two symbols d_i and d_j, the entry $C(d_i, d_j) = Prob(true\_value = d_i \mid observed\_value = d_j)$ is the conditional probability that d_i is the true value given that d_j is observed.

Figure 5.2 shows an example of the compatibility matrix between 5 symbols d_1, d_2, d_3, d_4, and d_5. An entry $C(d_i, d_j) > 0$ indicates that d_i might be (mis)represented as d_j in the observation; while $C(d_i, d_j) = 0$ implies that the symbol d_i cannot be represented as d_j despite the presence of noise. For instance, $C(d_1, d_2) = 0.1$ and $C(d_1, d_3) = 0$ in Figure 5.2. This means that there is a chance that a d_1 flips to a d_2 in the observation but it is impossible that a d_1 may turn to a d_3. Note that the compatibility is not necessary a symmetric measurement in the sense that $C(d_i, d_j) \neq C(d_j, d_i)$ may be true in some occasion. In Figure 5.2, $C(d_1, d_2) = 0.1$ and $C(d_2, d_1) = 0.05$. We also want to point out that, in the case where $C(d_i, d_i) < 1$, an observed symbol d_i does not always imply that d_i really occurs. $C(d_1, d_1) = 0.9$ implies that an observed d_1 truly represents itself with 90% probability and is a misrepresentation of some other symbol with 10% chance. (According to Figure 5.2, an observed d_1 has a 5% chance to be a misrepresentation of d_2 and d_3, respectively.) It is conceivable that (1) the compatibility matrix provides a meaningful measure to reveal the substance given the observation, and (2) the assessment of each entry has great impact to the final result. In practice, this matrix can be either given by a domain expert or learned from a training data set. In gene sequence analysis, this matrix can be obtained through clinical study[3]. In this chapter, we assume that the compatibility matrix is given by some domain expert in advance and will not elaborate on how to obtain and justify the value of each entry in the matrix. We also demonstrate in the experimental section that, even with a certain degree of error contained in the compatibility matrix, our model can still produce results of reasonable quality.

Given a pattern $P = d_1 d_2 \ldots d_l$ and an observed subsequence of l symbols $s = d'_1 d'_2 \ldots d'_l$, the conditional probability $Prob(P \mid s)$ represents the probability that s corresponds to an occurrence of P, and can be used as the indication of how much the pattern tallies with the observation. Therefore, we define the *match* of P in s to be the value of $Prob(P \mid s)$.

Definition Given a pattern $P = d_1 d_2 \ldots d_l$ and a subsequence of l observed symbols $s = d_1' d_2' \ldots d_l'$, the **match** of P in s (denoted by $M(P, s)$) is defined as the conditional probability $Prob(P \mid s)$.

Assuming that each observed symbol is generated independently, we have $M(P, s) = Prob(P \mid s) = \Pi_{1 \leq i \leq l} C(d_i, d_i')$. If $M(P, s) > 0$, then s is regarded as a (degraded) occurrence of P and $M(P, s)$ is viewed as the degree of which the pattern P is retained/reflected in s. We also say that P **matches** s if $M(P, s) > 0$ and P **does not match** s otherwise. For example, the match of $P_1 = d_1 d_2$ in a subsequence $s = d_1 d_3$ is $M(P_1, s) = C(d_1, d_1) \times C(d_2, d_3) = 0.9 \times 0.05 = 0.045$. However, the pattern $P_2 = d_1 d_5$ does not match s because $M(P_2, s) = C(d_1, d_1) \times C(d_2, d_5) = 0.9 \times 0 = 0$.

Definition For a symbol sequence S of length l_S and a pattern P of length l_P where $l_S \geq l_P$, the **match** of P in S is defined as the maximal match of P in every distinct subsequence (of length l_P) in S. That is, $M(P, S) = \max_{s \in S} M(P, s)$ where s is a subsequence of length l_P in S.

There are as many as $\begin{pmatrix} l_S \\ l_P \end{pmatrix}$ distinct subsequences[4] (of length l_P) in S. For example, there are 11 distinct subsequences of length 2 in $d_1 d_2 d_2 d_3 d_4 d_1$. The match of $P = d_1 d_2$ in this sequence is equal to $\max\{M(P, d_1 d_2), M(P, d_1 d_3),$ $M(P, d_1 d_4), M(P, d_1 d_1), M(P, d_2 d_2), M(P, d_2 d_3), M(P, d_2 d_4), M(P, d_2 d_1),$ $M(P, d_3 d_4), M(P, d_3 d_1), M(P, d_4 d_1)\} = \max\{0.72, 0.045, 0.09, 0, 0.09, 0.08,$ $0.005, 0.01, 0.005, 0, 0, 0\} = 0.72$. In fact, it is not necessary to probe every distinct subsequence in order to calculate the match of a pattern in a sequence. A dynamic programming approach can be used to compute the match in $O(l_P \times l_S)$ time. Given a pattern $P = d_1 d_2 \ldots d_{l_P}$ and a sequence $S = S_1 S_2 \ldots S_{l_S}$, the match $M(P, S) = M(d_1 d_2 \ldots d_{l_P}, S_1 S_2 \ldots S_{l_S})$ where

$$M(d_1 d_2 \ldots d_i, S_1 S_2 \ldots S_j) =$$
$$\max \begin{cases} M(d_1 d_2 \ldots d_{i-1}, S_1 S_2 \ldots S_{j-1}) \times C(d_i, S_j) \\ M(d_1 d_2 \ldots d_i, S_1 S_2 \ldots S_{j-1}) \end{cases} \tag{5.1}$$

where $M(\emptyset, \emptyset) = 1$. The process of calculating the match of the pattern $d_1 d_2$ in the sequence $d_1 d_2 d_2 d_3 d_4 d_1$ is shown in Figure 5.4.

Since the compatibility matrix is usually a sparse matrix, we can easily obtain an even more efficient algorithm to compute the match in nearly $\Theta(l_S)$ time [3]. Due to the space limitations, we will not elaborate on it in this chapter. Informally, the match of a pattern P in a sequence S is equal to the match of P in the subsequence (of S) which "best" aligns with P, and can be regarded

Sequence

		d1	d3	d2	d3	d4	d1
pattern	d1	0.9	0.9	0.9	0.9	0.9	0.9
	d2		0.045	0.72	0.72	0.72	0.72

Figure 5.4.   Dynamic Programming Method to Compute the Match

as an indicator of the degree of which the pattern P exhibits in the sequence S. We also say that P **matches** S if $M(P, S) > 0$ and P **does not match** S otherwise.

Definition Given a pattern P and a database D of N sequences, the **match** of P in D is the average match of P in every sequence in D, i.e. $M(P, D) = \frac{\Sigma_{S \in D} M(P, S)}{N}$.

Similar to the traditional support model, a user is asked to specify a minimum match threshold min_match to qualify significant patterns. All patterns that meet the min_match threshold are then referred to as **frequent** patterns. It is clear that the *match* model can accommodate misrepresentation of symbols due to noise in a seamless manner and provide a powerful means to properly separate the noise and change of behavior. Given a pattern $P = d_1 d_2 \ldots d_l$ and a subsequence $s = d'_1 d'_2 \ldots d'_l$, if the symbol at a given position (e.g., d'_i) in s cannot be a misrepresentation of the corresponding symbol (e.g., d_i) in P (i.e., $C(d'_i, d_i) = 0$), then the match of the pattern P in the subsequence s is 0 and the s would not be considered an occurrence of P. The match model also provides a natural bridge towards the traditional support model that does not allow partial match between pattern and data. In a noise-free environment, the conditional probability matrix becomes an *identity matrix* (i.e., $C(d_i, d_j)$ is 1 if $i = j$ and is 0 otherwise). The occurrence of a pattern becomes binary: either 1 (present) or 0 (absent). The match of a pattern in the data would be identical to the support of the pattern. In general, the more noise the environment assumes, the less skew the conditional probability distribution. Consider an extreme case where the sequence database is dominated by noise and no dependency exists between the observation and the true value. Then, all entries in the compatibility matrix would have the same value $\frac{1}{m}$ where m is the number of distinct symbols. As a result, all patterns would have exactly the same match value. This coincides with our intuition in the sense that, if the observed data is totally independent of the underlying system behavior, then no pattern should be considered more significant than others.

Figure 5.5(a) shows a database of 4 sequences. Figure 5.5(b) and (c) show the comparisons of supports and matches of each symbol and each pattern with two symbols, respectively. The number of patterns with positive match is usu-

ID	sequence
1	d1 d2 d3 d1
2	d4 d2 d1
3	d3 d4 d2 d1
4	d2 d2

(a) a sequence database

symbol	support	match
d1	0.75	0.538
d2	1.00	0.800
d3	0.50	0.400
d4	0.50	0.425
d5	0	0.075

(b) support and match of each symbol

pattern	support	match	pattern	support	match
d1 d1	0.25	0.250	d3 d4	0.25	0.136
d1 d2	0.25	0.203	d3 d5	0	0
d1 d3	0.25	0.160	d4 d1	0.50	0.363
d1 d4	0	0.025	d4 d2	0.50	0.321
d1 d5	0	0.034	d4 d3	0	0.036
d2 d1	0.75	0.560	d4 d4	0	0.053
d2 d2	0.25	0.210	d4 d5	0	0.004
d2 d3	0.25	0.160	d5 d1	0	0.068
d2 d4	0	0.052	d5 d2	0	0.032
d2 d5	0	0.035	d5 d3	0	0.008
d3 d1	0.50	0.349	d5 d4	0	0.028
d3 d2	0.25	0.179	d5 d5	0	0
d3 d3	0	0.037			

(c) support and match of patterns of length 2

pattern	match	pattern	match	pattern	match	pattern	match
d1 d1	0.01	d2 d3	0	d3 d5	0	d5 d2	0
d1 d2	0.08	d2 d4	0.08	d4 d1	0.01	d5 d3	0
d1 d3	0	d2 d5	0	d4 d2	0.08	d5 d4	0
d1 d4	0.01	d3 d1	0	d4 d3	0	d5 d5	0
d1 d5	0	d3 d2	0	d4 d4	0.01		
d2 d1	0.08	d3 d3	0	d4 d5	0		
d2 d2	0.64	d3 d4	0	d5 d1	0		

(d) the match contributed to each pattern by an observation of "d2 d2"

Figure 5.5. Comparison of *support* and *match*

ally much larger than that with positive support. In particular, as the pattern length increases, the match decreases at a much slower pace than the support. In the previous example (Figure 5.5(a)), consider patterns d_3, $d_3 d_2$, $d_3 d_2 d_2$, and $d_3 d_2 d_2 d_1$. Their supports are 0.5, 0.25, 0, 0, respectively; whereas their matches are 0.4, 0.179, 0.016, and 0.00522, respectively. This phenomenon is a direct consequence of the allowance of partial match between pattern and subsequence in the data. While each subsequence appearing in the data may increase the support of only one pattern with a full credit, its effect is dispersed among multiple patterns in terms of lifting their matches by various degrees. Figure 5.5(d) shows the amount of match that the subsequence $d_2 d_2$ may contribute to each pattern. There are totally 9 patterns that actually "benefit" from it. Note that the summation of these 9 numbers is still 1. It can be viewed as a "redistribution" of certain portion of the support in such a manner that the uncertainty introduced by noise is properly taken into account. For each pattern, the differential between the match and the support is the necessary rectification made towards the significance of the pattern. While the support can be viewed as the "face value" of a pattern, the match indeed represents the "expected value" (if no noise had presented).

The well-known Apriori property also holds on the match metric, which can be stated as in the following claims.

CLAIM 1.1 *The match of a pattern P in a symbol sequence S is less than or equal to the match of any subpattern of P in S.*

The proof of the above claim can be sketched as follows. Let $P = d_1 d_2 \ldots d_l$ and $P' = d'_1 d'_2 \ldots d'_{l'}$ be two patterns and P' is a subpattern of P ($l' \leq l$). Without loss of generality, assume that P' is a prefix of P (i.e., $d'_1 = d_1$, $d'_2 = d_2$, ..., $d'_{l'} = d_{l'}$). For any data subsequence $s = x_1 x_2 \ldots x_l$, the match of P in s is $M(P, s) = \Pi_{1 \leq i \leq l} C(d_i, x_i) \leq \Pi_{1 \leq i \leq l'} C(d_i, x_i)$ since $0 \leq C(d_i, x_i) \leq 1$ always holds. The match of P' in s is the maximum match between P' and any subsequence of length l' in s. Therefore, $M(P', s) \geq M(P', x_1 x_2 \ldots x_{l'}) = \Pi_{1 \leq i \leq l'} C(d'_i, x_i) = \Pi_{1 \leq i \leq l'} C(d_i, x_i)$ where $x_1 x_2 \ldots x_{l'}$ is a prefix of s. As a result, it must be true that $M(P', s) \geq M(P, s)$. By definition, the match of a pattern in a sequence is the maximal match of the pattern in every distinct subsequence of the sequence. It is very straightforward that, for any symbol sequence S, $M(P', S) \geq M(P, S)$ is also true. As a direct corollary of Claim 1.1, the follow claim also holds.

CLAIM 1.2 **(Apriori property)** *The match of a pattern P in a sequence database D is less than or equal to the match of any subpattern of P in D.*

A direct implication of the Apriori property is that, given a min_match threshold, the set of frequent patterns occupy a "contiguous portion" in the pattern lattice, and can be described using the notion of **border** [17]. Intuitively, the border demarcates the separation between the set of frequent patterns and the rest of the lattice, and can be represented by the set of frequent patterns whose immediate super-patterns are all infrequent. For example, if the patterns with solid circles are frequent in Figure 5.3, then the border should consist of three patterns: $d_1 d_2 d_3$, $d_1 d_2 d_5$, and $d_1 d_4$. These three patterns are also referred to as **border elements**. We sometimes use the phrase "the border of $match'$" as the abbreviation of "the border of frequent patterns given $match'$ as the min_match threshold".

An interesting observation is that, given a reasonable threshold, the number of frequent patterns at each level (in the super-/sub-pattern lattice) using the match metric is usually larger than that using the support. This is because, as the pattern length increases, the match decreases at a much slower pace than the support. Even though any algorithm powered (sometimes implicitly) by the Apriori property can be adopted to mine frequent patterns according to the match metric, it will produce a less efficient solution. The weakness becomes more substantial in mining sequence data since the length of a pattern can easily range up to dozens and even hundreds in many applications, such as biological expression. Even a direct generalization of previously proposed approach for mining long patterns under the support model (e.g., Max-Miner [4]) still requires many scans of the sequence database if the database is disk-resident. In the next section, we design a novel algorithm that can efficiently generate the border of frequent patterns in a few scans of the sequence database with very high confidence statistically.

A Border Collapsing Approach

For a given sequence database, we want to find patterns whose match satisfies a user-specified threshold min_match. To reduce the number of necessary passes through the input sequences, we propose a fast mining algorithm that can discover the border of frequent patterns in a few scans of the sequence database. Sampling technique is used to obtain a quick estimation of the border and additional scan(s) of the sequence database can be performed to finalize the border.

In order to obtain an estimation of the border of frequent patterns without examining the entire sequence database, we use the additive Chernoff bound [6, 14] to estimate the range of the match of a pattern from a sample of the data with a high statistical confidence (e.g., 99.99%). Let X be a random variable whose spread[5] is R. For example, in the context of the match model, the match can vary from 0 to 1 and therefore $R = 1$. Suppose that we have n independent observations of X, and the mean is μ. The Chernoff Bound states that with probability $1 - \delta$, the true mean of X is at least $\mu - \epsilon$, where

$$\epsilon = \sqrt{\frac{R^2 \ln(1/\delta)}{2n}}$$

For example, assume that the spread of a random variable is 1 and μ is the mean of 10000 samples of the random variable. Then we are able to say that the true value of the random variable is at least $\mu - 0.0215$ with 99.99% confidence. Similarly, with probability $1 - \delta$, the expected value of variable X is at most $\mu + \epsilon$. This provides the opportunity to estimate the range of the match for each pattern from a set of samples.

CLAIM 1.3 **(Chernoff bound estimation)** *Given a set of sample data and a threshold min_match, a pattern is* **frequent** *with probability $1 - \delta$ if $\mu_{match} > min\_match + \epsilon$ and is* **infrequent** *with probability $1 - \delta$ if $\mu_{match} < min\_match - \epsilon$, where μ_{match} is the match of the pattern in the sample data[6]. Those patterns (referred to as* **ambiguous** *patterns) whose matches in the sample are between $min\_match - \epsilon$ and $min\_match + \epsilon$ remain undecided and need further examination.*

An attractive property of the Chernoff bound is that it is independent of the probability distribution that generates the observations, as far as such probability distribution remains static during the entire process. This distribution-free nature is very important because the underlying distribution that characterizes the match of a pattern is usually unknown. However, this generality comes with the price of a more conservative bound than a distribution-dependent estimation and would require a larger number of observations to reach the same bound. This weakness sometimes prevents us from obtaining a tight bound when the sample size n is limited (e.g., due to memory size). Clearly, the number of ambiguous patterns highly depends on the value of ϵ which itself is a function of the sample size. A large number of ambiguous patterns may incur many scans of the entire sequence database. Therefore, the value of ϵ should be as small as possible. In order to further reduce ϵ under the constraint of memory capacity, instead of using $R = 1$, we employ an additional step to derive a more restricted spread R for the match of each pattern. According to the Apriori property (Claim 1.2), the match of a pattern is always less than or equal to the minimum match of each symbol in the pattern.

CLAIM 1.4 **(Restricted spread)** *The restricted spread R for the match of a pattern $d_1 d_2 \ldots d_l$ is $R = \min_{1 \le i \le l} match[d_i]$ where $match[d_i]$ is the match of the symbol d_i in the entire sequence database.*

For example, the match of $d_1 d_2$ in a sequence database would not exceed the minimum match of d_1 and d_2 in the database. If the matches of d_1 and d_2 are 0.1 and 0.05 in the database respectively, then the match of $d_1 d_2$ has to be between 0 and 0.05 (instead of the original spread 1) in the database. Thus, we can use $R = 0.05$ when applying the Chernoff bound and reduce the value of ϵ by 95%. (Note that ϵ is linearly proportional to R.) Therefore, before we examine the in-memory sample, a scan of the entire sequence database is performed to compute the match of each individual symbol. Note that as a by-product of this step, a random sample of the data can be easily obtained and kept in memory without any extra overhead. This sample set can then be used directly to classify patterns using Chernoff bound.

Nevertheless, when the pattern is long (e.g., in the range of dozens to hundreds of symbols) and the tolerable error is very small, the number of ambiguous patterns can be still considerably large and may require significant amount of computation and many scans through the database. This problem is more severe when the match (rather than the support) is used as the metric. To address this issue, we propose a **border collapsing** technique to ensure a minimum number of scans through the sequence database. Hence, the following three-fold algorithm is developed for mining the obscure patterns of length l.

1 While scanning the sequence database, find the match of each individual symbol and take a random sample of sequences.

2 Identify the borders that embrace the set of ambiguous patterns (i.e., whose match is between $min\_match - \epsilon$ and $min\_match + \epsilon$) using Chernoff bound based on the sample taken at the previous step.

3 Locate the border of frequent patterns in the entire sequence database via border collapsing.

A question one may concern is that, since the Chernoff bound only provides a probabilistic bound (rather than an absolute one), there is a small chance (bounded by δ) that a pattern P is frequent (i.e., its actual match in the entire sequence database is at least min_match) but P's match in the sample data is below $min\_match - \epsilon$. Even though the measured error is much smaller than δ in practice, it is important to understand the characteristic of these misclassified patterns. According to the above algorithm, P will be mislabeled as infrequent in the second phase. We now explore the impact of such mislabeled patterns to the quality of the result. Intuitively, it would be a less serious issue if the actual match of a mislabeled pattern is very close to $min\_match - \epsilon$ than the scenario where the actual match is far above min_match. The rationale is that, in the former case, one can always lower the threshold slightly to include the originally mislabeled patterns in the result. Therefore, the match distribution of mislabeled patterns is very important. Let $dis(P)$ be the difference between the actual match of a mislabeled pattern P and min_match. It is easy to derive from the Chernoff bound that the probability $Prob(dis(P) > \rho)$ diminishes exponentially as ρ grows. For example, $Prob(dis(P) > 2\rho) = Prob(dis(P) > \rho)^4$. This theoretically guarantees that the matches of most mislabeled patterns locate close to $min\_match - \epsilon$. This observation is also confirmed in the experimental results in the experimental study section. We now investigate each step in detail in the following subsections.

Phase 1: Finding Match of Individual Symbols and Sampling

In this phase, with one scan of the sequence database, we need to calculate the match of every symbol and obtain a sample set. Let's first look at the computation of the match of each symbol. A counter $match[d]$ is used for each distinct symbol $d \in \Theta$ to track the match value of d in the database. As we scan through each sequence D_i in the database, the match of d in D_i is $max\_match[d] = \max_{d' \in D_i} C(d, d')$. The value of $match$ of each symbol after examining each sequence in Figure 5.5(a) is shown in Figure 5.6. After we examine the entire sequence database, $match[d]$ holds match of each symbol d and d is a frequent symbol if $match[d] \geq min\_match$.

Obtaining the set of frequent symbols can be beneficial in two aspects. (1) According to the Apriori property, only frequent symbols may participate in a frequent pattern. With the set of frequent symbols on hand, we can eliminate unnecessary counters to a large extent. This is even more important to the

match	initial	sequence			
		1	2	3	4
d1	0	0.225	0.45	0.675	0.538
d2	0	0.2	0.4	0.6	0.8
d3	0	0.175	0.213	0.388	0.4
d4	0	0.025	0.213	0.4	0.425
d5	0	0.038	0.038	0.075	0.075

Figure 5.6.   Calculate match of each symbol in Figure 5.5(a)

match model since an occurrence of a symbol combination may trigger updates to match counters of multiple patterns. (2) The match of each (frequent) symbol in a (candidate) pattern can be used to provide a much restricted spread R of the match for this pattern to produce a much tighter bound ϵ. It will eliminate a large number of ambiguous patterns that need to be re-examined against the entire sequence database.

The computational complexity of this procedure is $O(N \times \overline{l_S} \times m)$ where $\overline{l_S}$ and m are the average sequence length and the number of distinct symbols, respectively. In the case where $\overline{l_S} \gg m$, it is easily to reduce the bound to $O(N \times (\overline{l_S} + m^2))$ by a simple optimization [27]. In summary, the computational complexity is $O(N \times \min\{\overline{l_S} \times m, \overline{l_S} + m^2\})$.

During the scan of the sequence database, a sample of sequences is also taken and stored in memory. Let n be the number of samples that can be held in memory. A very simple way [26] to guarantee a random sampling is to generate an independent uniform random variable for each sequence to determine whether that sequence should be chosen. At the beginning, a sequence will be chosen with probability $\frac{n}{N}$. Subsequently, if j sequences have been chosen from the first i sequences, then the next sequence will be chosen with probability $\frac{n-j}{N-i}$. The computational complexity of the sampling procedure is $O(N + n \times \overline{l_S})$, which makes the total computational complexity still $O(N \times \min\{\overline{l_S} \times m, \overline{l_S} + m^2\})$.

Phase 2: Ambiguous Pattern Discovery on Samples

Based on the samples taken in the previous phase, all patterns can be classified into three categories: *frequent* patterns, *infrequent* patterns, and *ambiguous* patterns, according to their observed matches in the sample data. In this phase, we want to find the two borders in the super-pattern/sub-pattern lattice, which separate these three categories. The border (denoted by FQT) between the frequent patterns and the ambiguous patterns is the set of frequent patterns whose immediate superpatterns are either ambiguous or infrequent, whereas the border (denoted by INFQT) between the ambiguous patterns and the infrequent patterns are the set of ambiguous patterns whose superpatterns are all infrequent. More specifically, these two borders correspond to the match thresholds

$min\_match + \epsilon$ and $min\_match - \epsilon$ respectively (with respect to the sample data).

Since the Apriori property holds on the match metric, many (border discovery) algorithms presented for mining frequent patterns (with respect to a support threshold) [4, 12, 17] can be adopted to solve this problem with one modification — the routine to update match(es) when examining each sample sequence. Let $P = d_1 d_2 \ldots d_l$ be a candidate pattern and $match[d_1, d_2, \ldots, d_l]$ denote the counter storing the match of the pattern $d_1 d_2 \ldots d_l$ in the sample data. By definition, the match of a pattern in the sample data is the average match of the pattern in every sample sequence. Equation 5.1 can be used to compute the match of a pattern $P = d_1 d_2 \ldots d_l$ in a sequence S.

With zero as the initial value, a straightforward way to compute the match of P in the sample is to accumulate the value of $match[d_1, d_2, \ldots, d_l]$ by an amount of $\frac{M(P,S)}{n}$ for each sample sequence S where $M(P, S)$ is the match of $P = d_1, d_2, \ldots, d_l$ in S. After we obtain $match[d_1, d_2, \ldots, d_l]$, P is labeled as

- a *frequent* pattern if $match[d_1, d_2, \ldots, d_l] > min\_match + \epsilon$;

- an *ambiguous* pattern if $match[d_1, d_2, \ldots, d_l] \in$
  $(min\_match - \epsilon, min\_match + \epsilon)$;

- an *infrequent* pattern otherwise;

where $\epsilon = \sqrt{\frac{R^2 \ln(1/\delta)}{2n}}$ and $R = \min_{1 \leq i \leq l} match[d_i]$. Since the sample data is in memory, any pruning technique (such as breadth-first, depth-first, looking-ahead, etc., [4, 12, 17]) may be used to locate the two borders FQT and INFQT that separate frequent patterns, ambiguous patterns, and infrequent patterns.

The optimal value of the confidence parameter δ used in the Chernoff bound is application dependent and can be adjusted by the user. Since the Chernoff bound is a very conservative bound, the actual error is usually much smaller than the theoretical probability δ. Empirically, when the pattern length is relatively short, a moderate value of δ (e.g., 0.001) is able to produce considerably high accuracy. This observation is also confirmed by our experiments discussed in the next section. However, as the pattern length grows, the number of patterns that need to be further verified against the entire database grows in an exponential pace. We will continue to investigate in this matter in the next section.

Assume the maximum length of any frequent pattern is $\widehat{l_P}$. There are up to $O(m^{\widehat{l_P}})$ distinct patterns of length up to $\widehat{l_P}$, where m is the number of symbols in the alphabet. The computational complexity of this phase is $O(m^{\widehat{l_P}} \times |S| \times \widehat{l_P} \times n)$ since it might take $O(|S| \times \widehat{l_P} \times n)$ computation to calculate the match of a pattern. Note that this only characterizes the theoretically worst scenario.

In practice, much less computation is usually required and all computation can be done efficiently as all sample data are in memory.

Phase 3: Border Collapsing

At this phase, we need to investigate ambiguous patterns further to determine the real border of frequent patterns. If the memory can hold the counters associated for all ambiguous patterns (i.e., all patterns between FQT and IN-FQT), a single scan of the entire sequence database would be able to calculate the exact match of each ambiguous pattern and the border of frequent patterns can be determined accordingly. However, we may experience with the scenario where a huge number of ambiguous patterns exist. This may occur when there are a large number of patterns whose matches happen to be very close to the threshold min_match, which is typically the case when the pattern is long. In such a case, multiple scans of the sequence database become inevitable.

Our goal of this phase is to efficiently collapse the gap between the two borders embracing the ambiguous patterns into one single border. An iterative "probing-and-collapsing" procedure can be employed. In order to minimize the expected number of scans through the database, the ambiguous patterns that can provide high collapsing effect are always probed first. A greedy algorithm can be developed to repeatedly choose the pattern with the most collapsing power among the remaining ambiguous patterns until the memory is filled up. A scan of the database is then performed to compute the matches of this set of patterns and the result is used to collapse the space of the remaining ambiguous patterns. This iterative process continues until no ambiguous pattern exist.

While the two borders embracing the ambiguous patterns act as the "floor" and the "ceiling" of the space of ambiguous patterns, an algorithm that is analogous to the binary search would serve our purpose. The patterns on the *halfway layer* between the two borders can provide the most collapsing effect and in turn should be probed first. The patterns on the quarterway layers are the set of patterns that can produce the most collapsing effect among the remaining ambiguous patterns, and so on. Consider the set of ambiguous patterns d_1, d_1d_2, $d_1d_2d_3$, $d_1d_2d_3d_4$, and $d_1d_2d_3d_4d_5$ in Figure 5.7(a). It is easy to see that $d_1d_2d_3$ has the most collapsing power. If $d_1d_2d_3$ is frequent, then d_1 and d_1d_2 must be frequent by the Apriori property. Otherwise (i.e., $d_1d_2d_3$ is infrequent), $d_1d_2d_3d_4$ and $d_1d_2d_3d_4d_5$ should be infrequent as well. Therefore, no matter whether $d_1d_2d_3$ is frequent or not, two other patterns (among the five) can be properly labeled without any further investigation on them. Similarly, we can justify that d_1d_2 and $d_1d_2d_3d_4$ have more collapsing power than the remaining two. In our algorithm, the patterns on the halfway layer (e.g., Layer 1 in Figure 5.7(a)), quarterway layers (e.g., Layers 2 and 3 in Figure 5.7(a)), $\frac{1}{8}$ layers, ... are identified successively until the memory is filled up by the corresponding counters. The process is carried out by sequentially computing the halfway layer between two adjacent layers calculated previously in a recursive manner. Given two layers of patterns, consider a pair of patterns P_1 and P_2 (one from each layer), where P_1 is a sub-pattern of P_2. The halfway patterns are the set of patterns that consist of $\lceil \frac{i_1+i_2}{2} \rceil$ symbols and are super-patterns of P_1 and sub-patterns of P_2, where i_1 and i_2 are the lengths of P_1 and P_2 respectively. Note that we do not physically store all ambiguous patterns. The set of ambiguous patterns that belong to the desired layer(s) are generated on the fly.

To better understand the effect brought by the border collapsing, let's assume that only patterns on the halfway layer are held in memory. If a halfway pattern turns out to be frequent, then all of its sub-patterns are frequent. Otherwise (i.e., the pattern is infrequent), all of its super-patterns are infrequent as well. In either case, one of these two borders is collapsed to that halfway pattern. For example, if we know that $d_1d_2d_3d_4d_5$ is on the border separating the ambiguous patterns and infrequent patterns while d_1 is on the border between frequent patterns and ambiguous patterns as shown in Figure 5.7(b). Thus, the patterns $d_1d_2d_3$, $d_1d_2d_4$, $d_1d_2d_5$, $d_1d_3d_4$, $d_1d_3d_5$, and $d_1d_4d_5$ are ambiguous patterns on the halfway layer between two borders and will be examined first. It is obvious that one of the borders would collapse to the halfway layer if these halfway patterns have homogeneous label (i.e., either all are frequent or all are infrequent). In this case, the space of ambiguous patterns is reduced by half. A more interesting scenario is that the halfway patterns have mixed labels (i.e., some of them are frequent while the rest are not), which turns out to provide even more collapsing effect. Assume that $d_1d_2d_3$ and $d_1d_2d_5$ are frequent (marked with solid circles on the halfway layer) while the remaining one (indicated by dashed circles on the halfway layer) are not. By applying the Apriori property, d_1, d_1d_2, d_1d_3, and d_1d_5 should also be frequent. Similarly, $d_1d_2d_3d_4$, $d_1d_2d_3d_5$, $d_1d_2d_4d_5$, $d_1d_3d_4d_5$, and $d_1d_2d_3d_4d_5$ are all infrequent. Note that only d_1d_4 still remains ambiguous as indicated by a solid rectangle in Figure 5.7(b). In general, if the memory can hold all patterns up to the "$\frac{1}{x}$ layer", the space of ambiguous patterns can be at least narrowed to $\frac{1}{x}$ of the original one where x is a power of 2. As a result, **if it takes a level-wise search y scans of the sequence database, only $O(\log_x y)$ scans are necessary when the border collapsing technique is employed**.

In summary, this approach can greatly reduce the number of scans through the sequence database by only examining a "carefully-chosen" small subset of all outstanding ambiguous patterns. While the traditional level-wise evaluation of ambiguous patterns push the border of frequent patterns forward across the pattern lattice in a gradual fashion; the border collapsing technique employs a globally optimal order to examine ambiguous patterns to minimize the overall computation and the number of scans through the database. When the pattern is relatively short, border collapsing achieves a comparable performance as the level-wise search. However, when the pattern is long (as in the applications

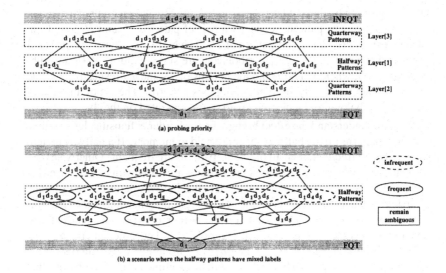

Figure 5.7.   Border Collapsing of Ambiguous Patterns

we addressed earlier in this chapter), the border collapsing technique can yield substantial improvement. We also want to mention that the proposed algorithm can also be used to mine long patterns with the support model efficiently.

Experimental Results for Obscure Patterns

Robustness of Match Model

Since misrepresentation of symbols may occur, some symbols may be substituted by others in the input sequence database. In this subsection, we compare the robustness of the support model and the match model with respect to varying degrees of noise. We use a protein database [20] that consists of 600K sequences of amino acids[7] as the *standard database* and generate *test databases* to do sensitivity analysis by embedding random noises. The objective here is not to evaluate the biological significance of the pattern discovery, but perform sensitivity analysis on noise level and other parameter values. A probability α is introduced to control the degree of noise. $\alpha = 0$ means no misrepresentation and a higher value of α implies a greater degree of misrepresentation. For each sequence S in the standard database, its counterpart S_t in the test database is generated as follows: for each amino acid d_i in the S, it will remain as d_i with probability $1 - \alpha$ and will be substituted by another amino acid d_j ($1 \leq j \leq m$ and $j \neq i$) with probability $\frac{\alpha}{m-1}$, where $m = 20$ is the number of distinct amino acids. S and S_t would have the same length. Each entry $C(d_i, d_j)$ in the corresponding compatibility matrix is $1 - \alpha$ if $i = j$ and is $\frac{\alpha}{m-1}$ otherwise. We also experienced with different noise distribution and, after a thorough study, we found that the degree of noise (rather than the distribution of the noise) plays a dominant role in the robustness of the model. Therefore, we only report the results under the assumption of uniform noise due to space limitations.

Let R_M be the set of patterns discovered via match model and R_S be the set of patterns discovered via support model on the standard sequence database with the same threshold $min\_match = min\_support = 0.001$. It is expected that $R_S = R_M$ since the match model is equivalent to the support model if no noise is assumed. This set of patterns will be used as the standard to justify the quality of the results generated from test database. Given a test database, let R'_M be the set of patterns discovered in the match model and R'_S be the set of discovered patterns under the support model. Figure 5.8(a) and (b) show the accuracy and completeness of these two models with respect to various degree of noise α, respectively. The accuracies of the match model and the support model are defined as $\frac{|R'_M \cap R_M|}{|R'_M|}$ and $\frac{|R'_S \cap R_S|}{|R'_S|}$ respectively. On the other hand, the completeness for the match and the support models are defined as $\frac{|R'_M \cap R_M|}{|R_M|}$ and $\frac{|R'_S \cap R_S|}{|R_S|}$, respectively. Intuitively, the accuracy describes how selective the model is while the completeness captures how well the model covers the expected results. For the match model, both the accuracy and the completeness are very high (i.e., more than 95%) due to the compensation of the compatibility matrices. This demonstrates that the match model is able to handle the noise in a proper manner. However, the support model appears vulnerable to the noise/misrepresentation in the data. When the misrepresentation factor α increases, the quality of the results by the support model degrades significantly. For example, when $\alpha = 0.6$, the accuracy and completeness of the support model are 61% and 33%, respectively.

With a given degree of noise (e.g., $\alpha = 0.1$), the accuracy and completeness of the support and match models with different pattern lengths are shown in Figure 5.8 (c) and (d), respectively. With longer pattern, the quality of the support degrades while the quality of the match model remains constant. This is due to the fact that for a long pattern, there is a higher probability that at least one position mutates.

We also experimented with the test database generated according to a noise level comparable to the actual amino acid mutations. This is done using the BLOSUM50 matrix [7] which is widely used to characterize the likelihood of mutations between amino acids in the computational biology community. We then use both the support and the match model to mine patterns on the test database with the minimum threshold set to 0.001. Comparing to the patterns discovered on the standard database, we found that both the accuracy and the completeness of the match model are well over 99% while the accuracy and the completeness of the support model are 70% and 50%, respectively.

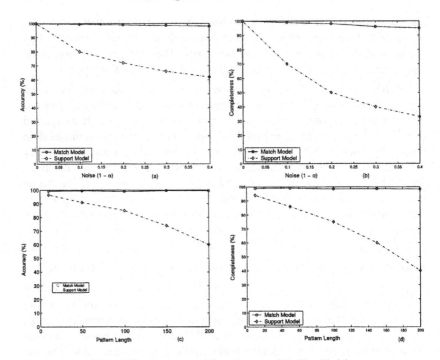

Figure 5.8.   Accuracy and Completeness of the Two Models

To further explore the importance of the match model, we build classifiers (e.g., decision trees) on proteins to predict their biological families using the set of patterns discovered under the match model and the support model, respectively. Each frequent pattern is regarded as a feature. A protein is considered to have a certain feature if the pattern appears in the protein. It is interesting to observe that the classifier corresponding to the match model is able to achieve over 85% correctness in predicting protein families while the classifier using the patterns discovered under the support model only reaches 53% correctness. We believe that this vast difference is due to the fact that the match model can successfully recover some vital features of proteins that fail to be captured by the support model if noise presents.

In the previous experiments, we assume that our knowledge of noise is "perfect", i.e., the compatibility matrix truly reflects the behavior of noise. However, in reality, the available compatibility matrix itself may contain some error and is indeed a (good) approximation of the real compatibility among symbols. This is typically the case when the compatibility matrix is generated from empirical studies. Thus, the quality of the compatibility matrix also plays a role in the performance of the match model. We also did some experiments to explore

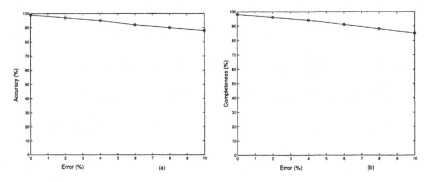

Figure 5.9. Robustness of Match Model

the robustness of the match model in this respect. Figure 5.9 shows the accuracy and completeness of the match model with respect to the amount of error contained in the compatibility matrix. In this experiment, we choose the test database generated from $\alpha = 0.2$. The error is incorporated into the compatibility matrix in the following manner. For each symbol d_i, the value of $C(d_i, d_i)$ is varied by $e\%$ (equally likely to be increased or decreased). The rest entries $C(d_j, d_i)$ $(j \neq i)$ in the same column are adjusted accordingly so that the summation $\Sigma_{1 \leq j \leq m} C(d_j, d_i)$ is still 1. Even though the completeness and accuracy degrades with the increase of error, the degradation is moderate even with high error rate. For example, with 10% error, our match model still can achieve 88% accuracy and 85% completeness. Note that the error in the compatibility matrix is usually very limited (i.e., $\ll 10\%$) in practice and hence the match model can perform very well.

Sample size

As mentioned previously, the sample size could affect the number of ambiguous patterns significantly and in turn impact the overall efficiency of our approach greatly. Figure 5.10 shows the number of ambiguous patterns with respect to the number of samples. The number of ambiguous patterns decrease significantly as a function of the number of samples. Also with greater degree of noise (i.e., larger α), the number of ambiguous patterns increases.

Spread of Match R

For any pattern, instead of applying the default value $R = 1$, a much constrained spread R can be estimated from the match of each involved symbol in the pattern and used to provide a tighter Chernoff bound. This leads to a significantly reduced number of ambiguous patterns. The same test database generated in the previous subsection are used here. Figure 5.11(a) shows the average

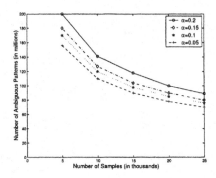

Figure 5.10.   Ambiguous Patterns w.r.t. Sample Size

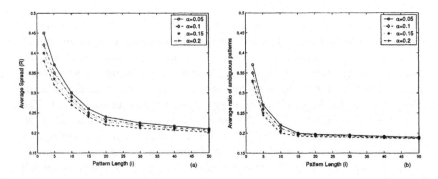

Figure 5.11.   Effects of Spread R

match spread R of a pattern with respect to the pattern length. R of a pattern is the minimum match of its involved symbols. Let $R(P)$ be the spread of the match of a pattern $P = d_1 d_2 \dots d_l$, then $R(P) = \min\{match(d_1), match(d_2), \dots, match(d_l)\}$. With longer pattern, the spread R becomes tighter. With higher degree of noise (i.e., larger value of α), the match spread reduces because the noise dilutes the strength of the true patterns. In Figure 5.11(b), we compute the ratio of the number of ambiguous patterns produced using the constrained R over that with the default $R = 1$. It is evident that the number of ambiguous patterns can be reduced to less than 20% (for pattern with more than 10 symbols) when the constrained R is applied. As a matter of fact, a five-folds pruning power is obtained.

Effects of Confidence $1 - \delta$

In the previous experiments, we fix confidence $1 - \delta$ as 0.9999. In this subsection, we are examining the effects of different δ. Figure 5.12 shows the

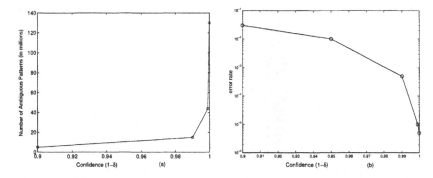

Figure 5.12. Effects of confidence $1 - \delta$

effect of $1 - \delta$ on the number of ambiguous patterns and the accuracy of the results. We assume that 200,00 samples are used for this test. With smaller confidence, the number of ambiguous patterns decreases dramatically because the error bound ϵ decreases, which implies a much faster response time. On the other hand, the error rate of the algorithm could increase slightly with a smaller confidence as shown in Figure 5.12(b). The error rate is defined as the ratio of the number of mislabeled patterns over the number of frequent patterns. However, since the Chernoff Bound is a distribution independent estimation, the bound that it provides is very conservative. The actual precision of the results is much higher than the specified confidence. For example, when confidence is 0.9, i.e., $\delta = 0.1$, the error rate is around 0.01. When $1 - \delta = 0.9999$, the error rate can diminish to the order of 10^{-6}.

Figure 5.13 shows the distribution of the matches of mislabeled patterns in the above experiment. It is clear that over 90% of the missed patterns are those whose real match is within 5% over the threshold, while no pattern missing whose real match is 15% over the threshold. This means that most missing patterns are very close to the threshold. This observation coincides with the theoretical analysis in the previous section.

Performance of Border Collapsing Algorithm

The overall efficiency of our border collapsing algorithm is demonstrated in Figure 5.14. Max-Miner [4] is one of the fastest algorithm for mining frequent long patterns, which employs a look-ahead technique. We adopt the Max-Miner as the deterministic algorithm to compare the performance. The only modification to the Max-Miner is the computation of match value of a pattern (instead of support value). Another algorithm we compared is the sampling based approach proposed by Toivonen [25]. In this approach, a level-wise search is used to finalize the border of frequent patterns after the sampling.

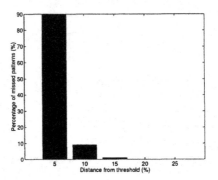

Figure 5.13.   Missed Patterns

We will refer to this approach as "sampling-based level-wise search" in the following discussion. The primary difference between this approach and our approach is that we employ a much more efficient method, namely border collapsing, to locate the border of frequent patterns. The confidence parameter of our algorithm is set to 0.9999. section) Figure 5.14(a) shows the CPU time of these three algorithms with respect to various match thresholds. Figure 5.14(b) shows the number of scans employed by these three algorithms.

It is evident that our algorithm can substantially reduce the CPU time and the number of scans through the database than both previous proposed schemes. This is due to the efficiency brought by the border collapsing technique. In our algorithm, the number of patterns that need to be examined against the entire database is much less than that in the other two algorithms. More specifically, when the match threshold is relatively high, our approach requires two scans of the sequence database while both Max-Miner and the sampling-based level-wise search requires at least five scans of data. As the match threshold decreases, the border collapsing algorithm requires three or four scans of the database while the other two approaches need 10 or more scans of the database. The significant reduction in number of database scans of our algorithm comes from the combined effect of sampling and border collapsing. We also would like to mention that the sampling-based level-wise approach spends majority of the time on finalizing the border of frequent patterns after estimating the border from the samples. We observed that there is a high likelihood that the final border is "far" from the estimated one and many scans of the data may be required before it is reached. This is because the match value usually changes very little from level to level in the pattern lattice especially when the pattern is long. This effect can clearly be observed from Figure 5.14(c).

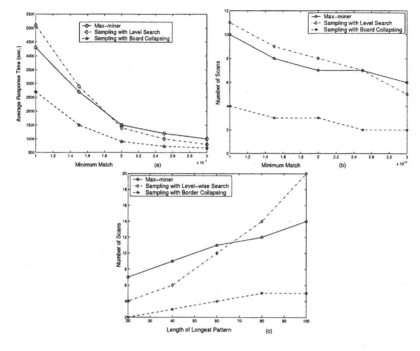

Figure 5.14. Performance of Three Algorithms

Scalability with respect to the Number of Distinct Symbols m

In all above experiments, we utilize the protein sequence data that consists of 20 symbols (i.e., amino acids). Now we analyze the performance of our algorithm with respect to the number of distinct symbols, m. In this experiment, we employ several synthetic data sets, each of which consists of 100K sequences and each sequence contains 1000 symbols on average. We vary the number of distinct symbols (m) in each data set. The minimum match threshold is set to 0.001. A compatibility matrix is constructed for each data set. In reality, most entries in a compatibility matrix is zero or near zero. Thus, the compatibility is generated in such a manner that a symbol is compatible to around 10% of other symbols with various degree. Figure 5.15(a)(b) shows the number of scans and response time of our algorithm, respectively. The number of scans decreases with the increase of m because less patterns are qualified to be significant. However, this trend does not hold for the response time. The average response time decreases initially, but increases when m gets large (e.g., greater than 10000). This is due to the fact that the size of the compatibility matrix is a quadratic function of m and the computation cost for each scan

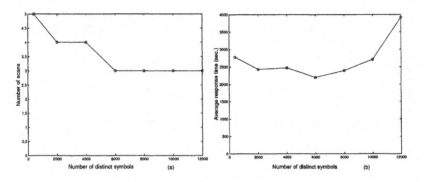

Figure 5.15.   Scalability w.r.t. the Number of Distinct Symbols

increases significantly. For example, if $m = 10000$, then it requires about 40MB space to store the compatibility matrix if each non-zero entry occupies 4 Bytes. The performance of our algorithm degrades when m is extremely large. Nevertheless, the algorithm performs very efficiently when the number of distinct symbols is within a reasonable range ($m \leq 10^4$).

2.   ApproxMap

Given a sequence database, a sequential pattern is a subsequence that appears frequently in the database. Since it has been proposed in [1], mining sequential patterns in large databases has become an important data mining task and has broad applications, such as business analysis, web mining, security, and bio-sequences analysis.

For example, supermarkets often collect customer purchase records in sequence databases. Sequential patterns in customer purchase database may indicate customers' buying habits and can be used to design promotion campaigns.

Most of the previous researches on sequential pattern mining adopt an *exact matching approach*. A pattern is supported by sequences in the database only if the pattern *exactly repeats* in the sequences. A sequence is regarded as a sequential pattern if the number of its occurrences in the sequence database passes a user specified support threshold. In general, two kinds of algorithms have been developed to find the complete set of sequential patterns.

First, the *appriori-like breadth-first search methods*, such as GSP [24] and SPADE [29], conduct a level-by-level candidate-generation-and-test pruning following the *Apriori property: any super-pattern of an infrequent pattern cannot be frequent*. In the first scan, they find the length-1 sequential patterns, i.e., single items frequent in the database. Then, length-2 candidates are assembled using length-1 sequential patterns. The sequence database is scanned the second time and length-2 sequential patterns are found. At each level, only potentially frequent candidates are generated and tested.

Second, the *projection-based depth-first search methods*, such as PrefixSpan [22], FreeSpan [13], and SPAM [2], avoid the costly candidate-generation-and-test operations by growing long patterns from short ones. Once a sequential pattern is found, all sequences containing that pattern are collected as a projected database. Local frequent items are found in the projected databases and used to extend the current pattern to longer ones.

Although sequential pattern mining has been extensively studied and many methods have been proposed, there are two inherent obstacles within the conventional framework.

- *Most methods mine sequential patterns with exact matching.* That is, a pattern gets support from a sequence in the database if and only if the pattern is exactly contained in the sequence. However, the exact matching approach often may not find general long patterns in the database. For example, many customers may share similar buying habits, but few of them follow an exactly same buying pattern. Thus, to find non-trivial interesting long patterns, we must consider mining *approximate* sequential patterns.

- *Most methods mine the complete set of sequential patterns.* When long patterns exist, mining the complete set of patterns is ineffective and inefficient. For example, if $\langle a_1 \cdots a_{20} \rangle$ is a sequential pattern, then each of its subsequence is also a sequential pattern. There are $(2^{20} - 1)$ patterns in total! On one hand, it is very hard for users to understand and manage a huge number of patterns. On the other hand, computing and storing a huge number of patterns is very expensive or even computationally prohibitive. In many situations, a user may just want the *long patterns* that cover many short ones.

Recently, mining compact expressions for frequent patterns, such as max-patterns [4] and frequent closed patterns [21], has been proposed and studied in the context of frequent itemset mining. However, mining max-sequential patterns or closed sequential patterns is far from trivial. Furthermore, in a noisy sequence database, the number of max- or closed sequential patterns still can be huge, and many of them are trivial for users.

In this section, we propose an effective and efficient framework for mining sequential patterns in databases of long sequences [16].

Problem Definition

Let $I = \{i_1, \ldots, i_l\}$ be a set of *items*. An *itemset* $X = \{i_{j_1}, \ldots, i_{j_k}\}$ is a subset of I. Conventionally, itemset $X = \{i_{j_1}, \ldots, i_{j_k}\}$ is also written as $(x_{j_1} \cdots x_{jk})$. A *sequence* $S = \langle X_1 \ldots X_n \rangle$ is an ordered list of itemsets, where $X_1, \ldots, X_n$ are all itemsets. A *sequence database* SDB is a multi-set of sequences.

A sequence $S_1 = \langle X_1 \cdots X_n \rangle$ is called a *subsequence* of sequence $S_2 = \langle Y_1 \cdots Y_m \rangle$, and S_2 a *super-sequence* of S_1, if $n \leq m$ and there exist integers $1 \leq i_1 < \cdots < i_n \leq m$ such that $X_j \subseteq Y_{i_j}$ $(1 \leq j \leq n)$. Given a sequence database SDB, the *support* of a sequence P, denoted as $sup(P)$, is the number of sequences in SDB that are super-sequences of P. Conventionally, a sequence P is called a *sequential pattern* if $sup(P) \geq min\_sup$, where min_sup is a user-specified *minimum support threshold*.

In many applications, people prefer long sequential patterns shared by many sequences. However, due to noise, it is very difficult to find a long sequential pattern exactly shared by many sequences. Instead, many sequences may *approximately* share a long sequential pattern.

Motivated by the above observation, we introduce the notion of *mining approximate sequential patterns*. Let $dist$ be a normalized distance measure of two sequences with domain $[0, 1]$. For sequences S, S_1 and S_2, if $dist(S, S_1) < dist(S, S_2)$, then S_1 is said be *more similar* to S than S_2 is.

Naïvely, we can extend the conventional sequential pattern mining framework to get an approximate sequential pattern mining framework as follows. Given a *minimum distance threshold* min_dist, the *approximate support* of a sequence P in a sequence database SDB is defined as $\widetilde{sup}(P) = \|\{S|(S \in SDB) \wedge (dist(S, P) \leq min\_dist)\}\|$. (Alternatively, the approximate support can be defined as $\widetilde{sup}(P) = \sum_{S \in SDB} dist(S, P)$. All the following discussion retains.) Given a minimum support threshold min_sup, all sequential patterns whose approximate supports passing the threshold can be mined.

Before we make any commitment, let us examine whether the above framework is good. Unfortunately, it may suffer from the following two problems.

First, *the mining may find many short and probably trivial patterns.* Short patterns tend to be easier to get similarity counts from the sequences than long patterns. Thus, short patterns may overwhelm the results.

Second, *the complete set of approximate sequential patterns may be larger than that of exact sequential patterns and thus difficult to understand.* By approximation, a user may want to get and understand the general trend and ignore the noise. However, a naïve output of the complete set of approximate patterns in the above framework may generate many (trivial) patterns and thus ruin the mining.

Based on the above analysis, we need to explore a more effective solution. We propose ApproxMAP, a *cluster and multiple alignment-based* approach, which works in two steps.

1 **Sequences in a database are clustered based on similarity**. Sequences in the same cluster may approximately follow some similar patterns.

2 **The longest approximate sequential pattern for each cluster is generated.** It is called the *consensus pattern*. To extract consensus patterns, a weighted sequence is derived for each cluster using multiple alignment to compress the sequential pattern information in the cluster. And then the longest consensus pattern best representing the cluster is generated from the weighted sequence.

Compared to the exact matching models and the naïve approximate pattern mining model, ApproxMAP has several distinct features.

1 ApproxMAP finds approximate sequential patterns from clusters based on multiple alignment. Even though a pattern is not exactly contained by many sequences, as long as it is shared by many sequences in the sense of multiple alignment, it will be identified by ApproxMAP.

2 In usual, each cluster has only one consensus pattern. Thus, many short and trivial patterns can be avoided. Since one cluster has multiple sequences, and one sequence only joins one cluster, the number of patterns mined by ApproxMAP is small.

3 ApproxMAP does not adopt a unified support threshold to prune patterns. Instead, it generates a pattern from each cluster regardless of its size. By doing so, ApproxMAP can find patterns strongly followed by a small number of sequences in the database. Such "rare" but "confident" patterns can be of great value in practice.

In the next two sections, we will discuss how to cluster sequences according to similarity and how to align sequences and generate patterns from clusters of sequences.

Clustering Sequences

In this section, we discuss *how to cluster sequences in a database*. First, let us develop a distance measure. In general, the *hierarchical edit distance* is commonly used as a distance measure for sequences. It is defined as the minimum cost of editing operations (i.e., insertions, deletions, and replacements) required to change one sequence to the other. An insertion operation on S_1 to change it towards S_2 is equivalent to a deletion operation on S_2 towards S_1. Thus, an insertion operation and a deletion operation have the same cost. We use $INDEL()$ to denote an insertion or deletion operation, and $REPL()$ to denote a replacement operation. Often, the following inequality is assumed.

$$REPL(X, Y) \leq INDEL(X) + INDEL(Y)$$

Given two sequences $S_1 = \langle X_1 \cdots X_n \rangle$ and $S_2 = \langle Y_1 \cdots Y_m \rangle$, the hierarchical edit distance between X and Y can be computed by dynamic programming using the following recurrence relation.

$$D(0,0)=0$$
$$D(i,0)=D(i-1,0)+INDEL(X_i) \text{ for } (1 \leq i \leq n)$$
$$D(0,j)=D(0,j-1)+INDEL(Y_j) \text{ for } (1 \leq j \leq m)$$
$$D(i,j)=\min \begin{cases} D(i-1,j)+INDEL(X_i) \\ D(i,j-1)+INDEL(Y_j) \\ D(i-1,j-1)+REPL(X_i,Y_j) \end{cases} \tag{5.2}$$
$$\text{for } (1 \leq i \leq n) \text{ and } (1 \leq j \leq m)$$

To make the edit distances comparable between sequences with various lengths, we normalize the results by dividing the hierarchical edit distance by the length of the longer sequence in the pair, and call it the *normalized edit distance*. That is,

$$dist(S_1, S_2) = \frac{D(n,m)}{\max\{\|S_1\|, \|S_2\|\}} \tag{5.3}$$

To make the hierarchical edit distance applicable to sequences of sets, we need to define the cost of edit operations (i.e., INDEL() and REPL() in Equation 5.2) properly. Here, we adopt the *normalized set difference* as the cost of replacement of sets.

$$\begin{aligned} REPL(X,Y) &= \frac{\|(X-Y)\cup(Y-X)\|}{\|X\|+\|Y\|} \\ &= \frac{\|X\|+\|Y\|-2\|X\cap Y\|}{\|X\|+\|Y\|} \end{aligned} \tag{5.4}$$

This measure has a nice property as follows.

$$0 \leq REPL() \leq 1$$

Moreover, it is a metric [5]. Following Equation 5.4, the cost of an insertion/deletion is

$$INDEL(X) = REPL(X, \emptyset) = 1,$$

where X is an itemset. Table 5.1 shows some examples on the calculation of normalized edit distances.

X	Y	$REPL(X,Y)$	X	Y	$REPL(X,Y)$
(a)	(a)	0	(a)	(b)	1
(a)	(ab)	$\frac{1}{3}$	(ab)	(cd)	1
(ab)	(ac)	$\frac{1}{2}$	(a)	$()$	1

Table 5.1.   Normalized edit distances between some itemsets.

Clearly, the normalized set difference is equivalent to the Sorensen coefficient.The Sorensen coefficient is defined as follows.

$$\begin{aligned} D_s(A,B) &= 1 - \frac{2\|A\cap B\|}{\|A-B\|+\|B-A\|+2\|A\cap B\|} \\ &= \frac{\|A\|+\|B\|-2\|A\cap B\|}{\|A\|+\|B\|} \\ &= REPL(A,B) \end{aligned}$$

It is also an index similar to the Jaccard coefficient. The Jaccard coefficient in dissimilarity notation is defined as follows.

$$\begin{aligned} D_J(A, B) &= 1 - \frac{\|A \cap B\|}{\|A \cup B\|} \\ &= 1 - \frac{\|A \cap B\|}{\|A - B\| + \|B - A\| + \|A \cap B\|} \end{aligned}$$

[19], except that it gives more weight to the common elements.

Using the hierarchical edit distance (Equation 5.3), we can apply a density-based clustering algorithm to cluster sequences. A density based clustering method groups the data according to the valley of the density function. The valleys can be considered as natural boundaries that separate the modes of the distribution [15, 8].

How can we define the density for a sequence? Intuitively, a sequence is *"dense"* if there are many sequences similar to it in the database. A sequence is *"sparse"*, or *"isolated"*, if it is not similar to any others, such as an outlier. Technically, the density of a sequence can be measured by a quotient of the number of similar sequences (nearest neighbors) against the space occupied by such similar sequences.

In particular, for each sequence S_i in a database S, let $d_1, \ldots, d_k$ be the k smallest non-zero values of $dist(S_i, S_j)$, where $S_j \neq S_i$, is a sequence in S. Then,

$$Density(S_i) = \frac{n}{\|S\|d^3}$$
$$\text{where} \quad d = \max\{d_1, \ldots, d_k\} \tag{5.5}$$
$$\text{and} \quad n = \|\{S_j \in S | dist(S_i, S_j) \leq d\}\|.$$

In Equation 5.5, n is the number of sequences in the k-nearest neighbor space (including all ties). Here k is a user-specified parameter. We adopt an algorithm from [9] as follows.

Input: *a set of sequences* $S = \{S_i\}$, *number of neighbor sequences* k;

Output: *a set of clusters* $\{C_j\}$, *where each cluster is a set of sequences;*

Method:

1 **Initialize every sequence as a cluster.**
   For each sequence S_i in cluster C_{S_i}, set $Density(C_{S_i}) = Density(S_i)$.

2 **Merge nearest neighbors based on the density of sequences.**
   For each sequence S_i, let $S_{i_1}, \ldots, S_{i_n}$ be the nearest neighbor of S_i, where n is defined in Equation 5.5. For each $S_j \in \{S_{i_1}, \ldots, S_{i_n}\}$, merge cluster C_{S_i} containing S_i with a cluster C_{S_j} containing S_j, if $Density(S_i) < Density(S_j)$ and there exists no S'_j having $dist(S_i, S'_j) < dist(S_i, S_j)$ and $Density(S_i) < Density(S'_j)$. Set the density of the new cluster to $\max\{Density(C_{S_i}), Density(C_{S_j})\}$.

3 **Merge based on the density of clusters**.

For all sequences S_i such that S_i has no nearest neighbor with density greater than that of S_i, but has some nearest neighbor, S_j, with density equal to that of S_i, merge the two clusters C_{S_j} and C_{S_i} containing each sequence if $Density(C_{S_j}) > Density(C_{S_i})$. This step is to merge "plateau neighbor regions"

It is easy to show that the above algorithm has complexity $O(kN_{seq})$. Intuitively, in the above algorithm, two clusters are merged if they are similar (in the sense of normalized edit distance). Thus, sequences in a resulting cluster are similar to each other.

The key parameter for the clustering process in the above algorithm is k, the number of nearest neighbors that the algorithm will search. A larger k value tends to merge more sequences, and results in a smaller number of large clusters, while a smaller k value tends to break up clusters. The benefit of using a small k value is that the algorithm can detect less frequent patterns. The tradeoff is that it may break up clusters representing strong patterns to generate multiple similar patterns. As shown in our performance study, in many applications, a value of k in the range from 3 to 10 works well.

Multiple Alignment and Pattern Generation

Once sequences are clustered by similarity, sequences within a cluster are similar to each other. Now, the problem becomes *how to summarize the general patterns in clusters and discover the trend*. In this section, we develop a method using multiple alignment. First, we discuss how to align sequences in a cluster, then we explore how to summarize sequences and generate patterns.

Multiple Alignment of Sequences

The global alignment of sequences is obtained by inserting empty itemsets (i.e., $\emptyset$) into sequences such that all the sequences have the same number of itemsets. The empty itemsets can be inserted into the front or the end of the sequences, or between any two consecutive itemsets [11].

The edit distance between two sequences S_1 and S_2 can be calculated by comparing itemsets in the aligned sequences one by one. If S_1 and S_2 have X and Y as their i^{th} aligned itemsets, respectively, where $(X \neq \emptyset)$ and $(Y \neq \emptyset)$, then a $REPL(X, Y)$ operation is required.

Otherwise, (i.e., S_1 and S_2 have X and $\emptyset$ as their i^{th} aligned itemsets, respectively) an $INDEL(X)$ operation is needed. The optimal alignment is the one in which the edit distance between the two sequences is minimized. Clearly, the optimal alignment between two sequences can be calculated by dynamic programming using Recurrence Relation 5.2.

In general, for a cluster C with n sequences $S_1, \ldots, S_n$, finding the *optimal global alignment* that minimizes

$$\sum_{j=1}^{n} \sum_{i=1}^{n} dist(S_i, S_j)$$

is an NP-hard problem [11], and thus is impractical for mining large sequence databases with many sequences. We need to look for some heuristics.

In a cluster, some sequences may be similar to many other sequences in the cluster. In other words, such sequences may have many close neighbors in terms of similarity. These sequences are most likely to be closer to the underlying patterns than the other sequences. It is more likely to get an alignment close to the optimal one, if we start the alignment with such "*seed*" sequences.

Intuitively, the *density* defined in Equation 5.5 measures the similarity between a sequence and its nearest neighbors. Thus, a sequence with a high density means that it has some neighbors very similar to it, and it is a good candidate for a "*seed*" sequence in the alignment. Based on the above observation, in ApproxMAP, we use the following heuristic to apply multiple alignment to sequences in a cluster.

HEURISTIC 2.1 *If sequences in a cluster C are aligned in a density-descending order, the alignment result tends to be good.*

As the first step in the clustering algorithm above, the density for each sequence is calculated. We only need to sort all sequences within a cluster in the density descending order.

How can we store the alignment results effectively? Ideally the result should be in a form such that we can easily align the next sequence to the current alignment. This will allow us to build a summary of the alignment step by step until all sequences in the cluster have been aligned. Furthermore, various parts of a general pattern may be shared with different strengths, i.e., some items are shared by more sequences and some by less sequences. The result should reflect the strengthes of items in the pattern.

Here, we propose a notion of weighted sequence as follows. A weighted sequence WS=$\langle X_1 : v_1, \ldots, X_l : v_l \rangle : n$ carries the following information:

1. the current alignment has n sequences, and n is called the *global weight* of the weighted sequence;

2. in the current alignment, v_i sequences have a non-empty itemset X_i aligned in the ith itemset, where $(1 \leq i \leq l)$;

3. an itemset in the alignment is in the form of $X_i = (x_{j_1} : w_{j_1}, \ldots, x_{j_m} : w_{j_m})$, which means, in the current alignment, there are w_{j_k} sequences that

S1 = <(ag)(f)(bc)(ae)(h)>   S3 = <(a)(b)(de)>   S5=<(bci)(de)>
S2 = <(ae)(h)(b)(d)>   S4 = <(a)(bcg)(d)>

Seq–id					
S1	<(ag)	(f)	(bc)	(ae)	(h)>
S2	<(ae)	(h)	(b)	(d)	>
S3	<(a)		(b)	(de)	>
S4	<(a)		(bcg)	(d)	>
S5	<		(bci)	(de)	>
W.S.	<(a:4,e:1,g:1):4 (f:1,h:1):2 (b:5,c:3,g:1,i:1):5 (a:1,d:4,e:3):5 (h:1):1>:5				

Figure 5.16.   Sequences in a Cluster and the Complete Alignment

have item x_{j_k} in the ith position of the alignment, where $(1 \leq i \leq l)$ and $(1 \leq k \leq m)$.

The notations and the ideas are demonstrated in the following example.

EXAMPLE 5.1 (MULTIPLE ALIGNMENT) Suppose that, in a cluster C, there are 5 sequences as shown in the left two columns in Figure 5.16. The density descending order of these sequences is S_3-S_2-S_4-S_5-S_1. The sequences are aligned as follows.

First, sequences S_3 and S_2 are aligned as shown in Figure 5.17.

$$S_3 \quad \langle(a) \qquad\qquad (b) \quad (de)\rangle$$
$$S_2 \quad \langle(ae) \qquad (h) \quad (b) \quad (d)\rangle$$
$$WS_1 \langle(a:2,e:1):2(h:1):1(b:2):2(d:2,e:1):2\rangle:2$$

Figure 5.17. S_3 and S_2 are aligned resulting in WS_1.

Here, we use a *weighted sequence* WS_1 to summarize and compress the information about the alignment. Since the first itemsets of S_3 and S_2, (a) and (ae), are aligned, the first itemset in the weighted sequence WS_1 is $(a:2,e:1):2$. It means that the two sequences are aligned in this itemset, and a and e appear twice and once, respectively. The second itemset in WS_1, $(h:1):1$, means there is only one sequence with an itemset aligned in this itemset, and item h appears once.   □

After the first step, we need to iteratively align other sequences with the current weighted sequence. The weighted sequence does not explicitly keep the information about various itemsets in the sequences. Instead, this information is summarized into the item weights in the weighted sequence. These weights need to be taken into account when aligning a sequence to a weighted sequence.

Thus, instead of using Equation 5.4 directly to calculate the distance between a weighted sequence and a sequence in the cluster, we adopt a *weighted replace cost* as follows.

Let $X = (x_1 : w_1, \ldots, x_m : w_m) : v$ be an itemset in a weighted sequence, while $Y = (y_1 \cdots y_l)$ is an itemset in a sequence in the database. Let n be the global weight of the weighted sequence. The replace cost is defined as

$$REPL(X, Y) = \frac{e_R \cdot v + n - v}{n}$$

$$\text{where} \quad e_R = \frac{\sum_{i=1}^{m} w_i + \|Y\| v - 2 \sum_{x_i \in Y} w_i}{\sum_{i=1}^{m} w_i + \|Y\| v}$$

(5.6)

Accordingly, we have $INDEL(X) = REPL(X, \emptyset) = 1$ and $INDEL(Y) = REPL(Y, \emptyset) = 1$.

The rationale of the Equation 5.6 is as follows. After aligning a sequence, its alignment information is incorporated into the weighted sequence. There are two cases.

- *A sequence may have a non-empty itemset aligned in this itemset.* Then, e_R is the estimated average replacement cost for all sequences that have a non-empty itemset aligned in this itemset. There are in total v such sequences.

- *A sequence may have an empty itemset aligned in this itemset.* Then, we need an $INDEL()$ operation (whose cost is 1) to change the sequence to the one currently being aligned. There are in total $(n - v)$ such sequences.

Equation 5.6 estimates the average of the cost in the two cases.

EXAMPLE 5.2 (MULTIPLE ALIGNMENT (CON'D)) In the next step, the weighted sequence WS_1 and the third sequence S_4 are aligned as shown in Figure 5.18. Similarly, we can align the remaining sequences. The results are shown in Figure 5.19.

WS_1	$\langle (a:2, e:1) : 2$	$(h:1):1$	$(b:2):2$	$(d:2, e:1):2 \rangle$	$:2$
S_4	$\langle (a)$		(bcg)	$(d) \rangle$	
WS_2	$\langle (a:3, e:1):3$	$(h:1):1$	$(b:3, c:1, g:1):3$	$(d:3, e:1):3 \rangle$	$:3$

Figure 5.18. Sequences WS_1 and S_4 are aligned.

The alignment result for all sequences are summarized in the weighted sequence WS_4 shown in Figure 5.19. After the alignment, we only need to store WS_4. All the sequences in the cluster are not needed any more in the remainder of the mining. □

Aligning the sequences in different order may result in slightly different weighted sequences. To illustrate the effect, Figure 5.20 shows the alignment result using the id order, S_1-S_2-S_3-S_4-S_5.

WS2	<(a:3, e:1):3	(h:1):1	(b:3,c:1,g:1):3	(d:3,e:1):3>	:3
S5	<		(bci)	(de)>	
WS3	<(a:3,e:1):3	(h:1):1	(b:4,c:2,g:1,i:1):4	(d:4,e:2):4>		:4
S1	<(ag)	(f)	(bc)	(ae)	(h)>	
WS4	<(a:4,e:1,g:1):4	(f:1,h:1):2	(b:5,c:3,g:1,i:1):5	(a:1,d:4,e:3):5	(h:1):1>	:5

Figure 5.19. The alignment of remaining sequences.

S1 = <(ag)(f)(bc)(ae)(h)>   S3 = <(a)(b)(de)>   S5=<(bci)(de)>
S2 = <(ae)(h)(b)(d)>   S4 = <(a)(bcg)(d)>

Seq–id					
S1	<(ag)	(f)	(bc)	(ae)	(h)>
S2	<(ae)	(h)	(b)		(d)>
S3	<(a)		(b)		(de)>
S4	<(a)		(bcg)		(d)>
S5	<		(bci)		(de)>
W.S.	<(a:4,e:1,g:1):4 (f:1,h:1):2 (b:5,c:3,g:1,i:1):5 (a:1,e:1):1 (d:4,e:2,h:1):5				

Figure 5.20. Aligning Sequences in Another Order

Interestingly, the two alignment results are quite similar, only some items shift positions slightly. This causes the item weights to be reduced slightly. As verified by our extensive empirical evaluation, the alignment order has little effect on the underlying patterns.

As shown in the above example, for a cluster of n sequences, the complexity of the multiple alignment of all sequences is $O(n \cdot t)$, where t is the maximal cost of aligning two sequences. The result of the multiple alignment is a weighted sequence. A weighted sequence records the information of the alignment. Once a weighted sequence is derived, the sequences in the cluster will not be visited anymore.

Now, the remaining problem is *how to generate patterns from weighted sequences*.

Generation of Consensus Patterns

As shown in Multiple Alignment Sequences section , a weighted sequence records the statistics of the alignment of the sequences in a cluster. Intuitively, a pattern can be generated by picking up parts of a weighted sequence shared by most sequences in the cluster.

For a weighted sequence $WS = \langle (x_{11} : w_{11}, \ldots, x_{1m_1} : w_{1m_1}) : v_1, \ldots, (x_{l1} : w_{l1}, \ldots, x_{lm_l} : w_{lm_l}) : v_l \rangle : n$, the *strength* of item $x_{ij} : w_{ij}$ in the ith

itemset is defined as $\frac{w_{ij}}{n} \cdot 100\%$. Clearly, an item with a larger strength value indicates that the item is shared by more sequences in the cluster.

Motivated by the above observation, a user can specify a *strength threshold* $min\_strength$ ($0 \leq min\_strength \leq 1$). A *consensus pattern* P can be extracted from a weighted sequence by removing items in the sequence whose strength values are lower than the threshold.

EXAMPLE 5.3 (CONSENSUS PATTERN GENERATION)
Suppose a user specifies a strength threshold $min\_strength = 30\%$. The consensus pattern extracted from weighted sequence WS_4 is $\langle (a)(bc)(de) \rangle$.

Interestingly, if we compare the sequences in the sequence database (Figure 5.16) and the consensus pattern mined from the database, the pattern is shared by the sequences, but it is not exactly contained in any one of them. In particular, every sequence except S_2 approximately contains the pattern by one insertion. These evidences strongly indicate that the consensus pattern is the general template behind the data.   □

Empirical Evaluations

In this section, we report an extensive set of empirical evaluations performed on ApproxMAP. We use both synthetic and real data sets to test the method. All experiments were run on a 4-processor Pentium III 700 MHz Linux machine with 2GB of main memory. Our program only uses one CPU in all experiments.

Synthetic Data Generator

To gain insight on how ApproxMAP behaves under various settings, we use the IBM synthetic data generator [1] to generate various synthetic data sets. The IBM data generator takes several parameters and outputs a sequence database as well as a set of base patterns. The sequences in the database are generated in two steps. First, *base patterns* are generated randomly according to the user's specification. Then, these base patterns are corrupted (drop random items) and merged to generate the sequences in the database. Thus, these base patterns are approximately shared by many sequences. The base patterns are the underlying template behind the database.

We summarize the parameters of the data generator and the mining in Table 5.2.

Evaluation Criteria

Although many people have used this synthetic data generator to generate benchmark data sets for sequential pattern mining, to the best of our knowledge, no previous study examines whether the sequential pattern mining can discover

Notation	Meaning	Default value
$\|I\|$	# of items	1000
N_{seq}	# of data sequences	10000
N_{pat}	# of base pattern sequences	100
L_{seq}	Avg. # of itemsets per data sequence	10
L_{pat}	Avg. # of itemsets per base pattern	$0.7 \cdot L_{seq}$
I_{seq}	Avg. # of items per itemset in the database	2.5
I_{pat}	Avg. # of items per itemset in base patterns	$0.7 \cdot I_{seq}$
k	# of neighbor sequences	5
$min\_strength$	The strength threshold consensus patterns	50%

Table 5.2.   Parameters and default values for the data generator and the mining.

the base patterns properly. It is difficult for conventional sequential pattern mining methods to uncover the long base patterns. Furthermore, the conventional methods usually generate much more than just the base patterns. To test the effectiveness of ApproxMAP, we examine whether ApproxMAP can find the base patterns without generating many trivial or irrelevant patterns.

Specifically, for a base pattern B and a consensus pattern P, we denote $B \otimes P$ as the longest common subsequence S of both B and P. Then, we define the *recoverability* as follows to measure the quality of the mining.

$$R = \sum_{\text{base pat } B} E(F_B) \cdot \min \left\{ \begin{array}{l} 1 \\ \dfrac{\max_{\text{con pat } P}\{\|B \otimes P\|\}}{E(L_B)} \end{array} \right. \tag{5.7}$$

where $E(F_B)$ and $E(L_B)$ are expected frequency and expected length of base pattern B. They are given by the data generator. Since $E(L_B)$ is an expected value, sometimes the actual observed value, $\max\{\|B \otimes P\|\}$ is greater than $E(L_B)$. In such cases, we cutoff the value of $\frac{\max\{\|B \otimes P\|\}}{E(L_B)}$ to be 1 so that recoverability stays between 0 and 1.

Intuitively, if the recoverability of the mining is high, major parts of the base patterns will be found.

Effectiveness of ApproxMAP

We ran many experiments with various synthetic data sets. The trend is clear and consistent. Limited by space, we report only the results on some selected data sets here.

First, let us take a close look at the mining result from a small data set with $1,000$ sequences. The data generator uses 10 base patterns to generate the data. ApproxMAP mines the data set using the following parameters: the number of nearest neighbors $k = 6$ (for clustering), and strength threshold

$min\_strength = 30\%$. Under such settings, ApproxMAP also finds 10 consensus patterns. The patterns are shown in Table 5.3.

Exp freq	Exp len	Observed pat. len	Type	Patterns
0.21	0.66	13	$ConPat_1$	$\langle(15,16,17,66)(15)(58,99)(2,74)(31,76)$ $(66)(62)\rangle$
		14	$BasePat_1$	*⟨(15,16,17,66)(15)(58,99)(2,74) (31,76)(66)(62)(93)⟩*
0.161	0.83	11	$ConPat_2$	$\langle(16,22)(29,99)(94)(45,67)(50)(96)(51)$ $(66)\rangle$
		13	$ConPat_3$	$\langle(22,50,66)(16)(29,99)(94)(45,67)$ $(12,28,36)(50)\rangle$
		19	$ConPat_4$	$\langle(22,50,66)(16)(29,99)(94)(45,67)$ $(12,28,36)(50)(96)(51)(66)(2,22,58)\rangle$
		22	$BasePat_2$	*⟨(22,50,66)(16)(29,99)(94)(45,67) (12,28,36)(50)(96)(51)(66)(2,22,58) (63,74,99)⟩*
0.141	0.82	11	$ConPat_5$	$\langle(22)(22)(58)(2,16,24,63)(24,65,93)(6)\rangle$
		14	$BasePat_3$	*⟨(22)(22)(58)(2,16,24,63)(24,65,93) (6)(11,15,74)⟩*
0.131	0.9	11	$ConPat_6$	$\langle(31,76)(58,66)(16,22,30)(16)(50,62,66)\rangle$
		15	$BasePat_4$	*⟨(31,76)(58,66)(16,22,30)(16) (50,62,66)(2,16,24,63)⟩*
0.123	0.81	13	$ConPat_7$	$\langle(43)(2,28,73)(96)(95)(2,74)(5)(2)$ $(24,63)(20)\rangle$
		14	$BasePat_5$	*⟨(43)(2,28,73)(96)(95)(2,74)(5)(2) (24,63)(20)(93)⟩*
0.121	0.77	8	$ConPat_8$	$\langle(63)(16)(2,22)(24)(22,50,66)\rangle$
		9	$BasePat_6$	*⟨(63)(16)(2,22)(24)(22,50,66)(50)⟩*
0.0539	0.6	11	$ConPat_9$	$\langle(70)(58,66)(22)(74)(22,41)(2,74)$ $(31,76)\rangle$
		13	$BasePat_7$	*⟨(70)(58,66)(22)(74)(22,41)(2,74) (31,76)(2,74)⟩*
0.0135	0.91	15	$ConPat_{10}$	$\langle(20,22,23,96)(50)(51,63)(58)(16)$ $(2,22)(50)(23,26,36)\rangle$
		17	$BasePat_8$	*⟨(20,22,23,96)(50)(51,63)(58)(16) (2,22)(50)(23,26,36)(10,74)⟩*
0.0382	0.78	7	$BasePat_9$	*⟨(88)(24,58,78)(22)(58)(96)⟩*
0.00809	0.66	17	$BasePat_{10}$	*⟨(16)(2,23,74,88)(24,63)(20,96)(91) (40,62)(15)(40)(29,40,99)⟩*

Table 5.3. Consensus patterns and the base patterns in a small data set.

As shown, each of the 10 consensus patterns match some base pattern. The consensus patterns do not cover the last two base patterns. The recoverability is 92.46%. In general, the consensus patterns recover major parts of the base patterns. The consensus patterns cannot recover the complete base patterns because, during the data generation, only parts of base patterns are embedded into a sequence. Hence, some items in the base patterns may have much lower frequency than the others. We also checked the worksheet of ApproxMAP. The less frequent items in the base patterns are in the weighted sequence. However, they are not included in the consensus patterns because their item strengths are weak.

It is interesting to note that a base pattern may be recovered by multiple consensus patterns. For example, ApproxMAP forms three clusters whose consensus patterns approximate base pattern $BasePat_2$. This is because $BasePat_2$ is long (the actual length of the base pattern is 22 items and the expected length of the pattern in the data is 18 items) and has a high expected frequency (16.1%). Therefore, many data sequences in the database are generated using $BasePat_2$ as a template. As discussed above, sequences are generated by removing various parts of the base pattern and inserting noise items. Thus, two sequences using the same long base pattern as the template are not necessarily similar to each other. As a consequence, the sequences generated from a long base pattern can be partitioned into multiple clusters by ApproxMAP.

In all the consensus patterns, there is only one item (in the first itemset of $ConPat_2$) that does not appear on the corresponding position in the base pattern. This fact indicates that the consensus patterns are highly shared by sequences in the database.

Based on the above analysis, we can see that ApproxMAP summarizes $1,000$ sequences in this small data set into 10 consensus patterns accurately. The 10 consensus patterns resemble the base patterns that generates the sequences very well (recoverability=92.46%). No trivial nor irrelevant pattern is generated.

Effects of Parameters And Scalability

Now, let us examine the effects of various parameters on the recoverability. The default configuration of the data sets used in the remaining experiments is given in Table 5.2. The expected frequencies of the 100 base patterns range from 7.63% to 0.005%. We test the recoverability against 5 factors, namely the nearest neighbor parameter k, the number of items in the data set $\|I\|$, the data set size in terms of number of sequences N_{seq}, the average number of itemsets in a sequence L_{seq}, and the average number of items per itemset I_{seq}.

First, we fix other settings and vary the value of k from 2 to 10, where k is the nearest neighbor parameter in the clustering. The results are shown in Figure 5.21. As analyzed before, a larger value of k produces less number of clusters, which leads to less number of patterns. As expected, as k increases in Figure 5.21(a), the number of consensus patterns decreases. This causes loss of some weak base patterns and thus the recoverability decreases slightly, as shown in

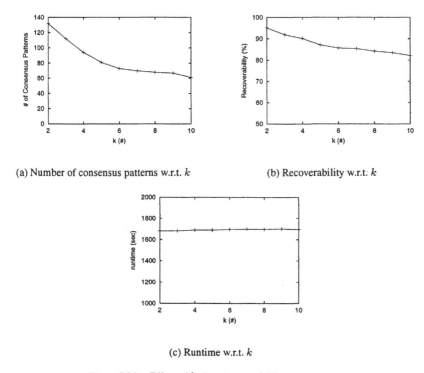

(a) Number of consensus patterns w.r.t. k (b) Recoverability w.r.t. k

(c) Runtime w.r.t. k

Figure 5.21.   Effect of k, the nearest neighbor parameter.

Figure 5.21(b). In addition, like most density based clustering algorithms, the recoverability is sustained for a range of k (5 to 9) (Figure 5.21(b)). Figure 5.21(c) indicates that the performance of ApproxMAP is not very sensitive to parameter k. It is stable.

Second, we studied the effect of the number of items in the set I, $\|I\|$. A smaller value of $\|I\|$ results in a denser database (i.e., patterns are with higher frequencies) because the items come from a smaller set of literals. In multiple alignment, the positions of the items have strong effect on the results. Thus, even when the density of the database changes, the alignment statistics does not change substantially. We observe that the performance of ApproxMAP in terms of number of consensus patterns, recoverability and runtime remains stable. Limited by space, we omit the details here.

Third, we test the effect of data set size in terms of number of sequences in the data set. The results are shown in Figure 5.22. As the data set size increases, the number of clusters also goes up. That increases the number of consensus patterns, as shown in Figure 5.22(a). However, it is interesting to note that the

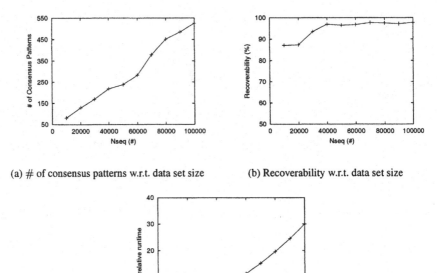

(a) # of consensus patterns w.r.t. data set size   (b) Recoverability w.r.t. data set size

(c) Runtime w.r.t. data set size

Figure 5.22.   Effect of N_{seq}, the number of sequences in the data set.

recoverability also increases as the data set size goes up, as shown in 5.22(b). It can be explained as follows.

With multiple alignment, the more the sequences in the data set, the easier the recovery of the base patterns. In large data sets, there are more sequences approximating the patterns. For example, if there are only 1000 sequences, a base pattern that occurs in 1% of the sequences will only have 10 sequences approximately similar to it. However, if there are 100, 000 sequences, then there would be 1, 000 sequences similar to the the base pattern. It would be much easier for ApproxMAP to detect the general trend from 1, 000 sequences than from 10 sequences. Moreover, we observe that ApproxMAP is scalable w.r.t. data set size, as shown in Figure 5.22(c).

We observe similar effects from factors L_{seq}, the average number of itemsets in a sequence, and I_{seq}, the average number of items per itemset in the sequences. Limited by space, we omit the details here.

The above analysis strongly indicates that ApproxMAP is effective and scalable in mining large databases with long sequences.

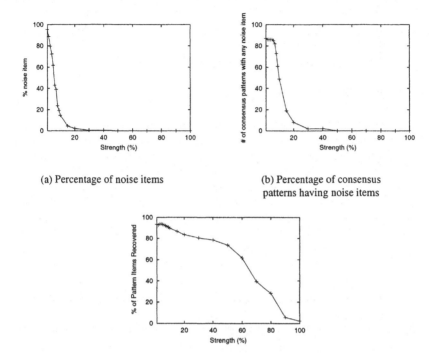

(a) Percentage of noise items

(b) Percentage of consensus
patterns having noise items

(c) Percentage of items in base patterns appearing in consensus patterns

Figure 5.23. Effects of strength thresholds

The Strength Threshold

In ApproxMAP, we use a strength threshold $min\_strength$ to filter out noise from weighted sequences. Here, we test the effect of the strength threshold on the mining results.

In Figure 5.23, the percentage of noise items, the percentage of consensus patterns having any noise item, and the percentage of items in base patterns appearing in consensus patterns w.r.t. the strength threshold is plotted, respectively. As $min\_strength$ is decreased from 100%, both the percentage of noise items in Figure 5.23(a) and the number of patterns containing any noise item in Figure 5.23(b) start to increase at $min\_strength = 30\%$ and increase drastically at 20%. This indicates that items with $strength \geq 30\%$ are probably items in the base patterns. On the other hand, the average percentage of items from the base pattern recovered by the consensus pattern is reasonably stable when $min\_strength \leq 50\%$, but goes down quickly when $min\_strength > 50\%$.

This observation is also verified by experiments on other data sets. Our experience with data sets generated by the IBM data generator indicates that 20%-50% is in fact a good range for $min\_strength$.

In all our experiments, the default value of $min\_strength$ is 50%.

Spurious Patterns

Spurious patterns are defined as consensus patterns that are not similar to any base pattern. These are in essence consensus patterns with almost all noise items. Our experimental results show that ApproxMAP is able to uncover the base patterns without generating many *spurious patterns*, i.e., the number of consensus patterns that are very different from all of the base patterns is usually very small. Almost all consensus patterns are very close to the base patterns. Only a very small number of noise items appear in the consensus patterns.

The Orders in Multiple Alignment

Now, we studied the sensitivity of the multiple alignment results to the order of sequences in the alignment. We compare the mining results using the density-descending order, density-ascending order, and the sequence-id ascending order. As expected, although the exact alignment changes slightly depending on the orders, it has very limited effect on the consensus patterns. All three orders generated the exact same number of patterns, which were very similar to each other. Therefore, the recoverability is basically identical.

This result shows that, while aligning patterns in density descending order tends to improve the alignment quality, ApproxMAP itself is robust w.r.t. alignment orders.

Case Study: Mining The Welfare Services Database

We also tested ApproxMAP on a real data set of welfare services accumulated over a few years in North Carolina State. The services have been recorded monthly for children who had a substantiated report of abuse and neglect, and were placed in foster care. There were 992 such sequences. In summary we found 15 interpretable and useful patterns.

In total 419 sequences were grouped together into one cluster which had the following consensus pattern.

$$\langle (RPT)(INV, FC)\overbrace{(FC)\cdots(FC)}^{11}\rangle$$

In the pattern, RPT stands for a report, INV stands for an investigation, and FC stands for a foster care service. The pattern indicates that many children

$$\langle (RPT)(INV, FC, T)(FC, T) \overbrace{(FC, HM) \cdots (FC, HM)}^{8}(FC)(FC, HM) \rangle$$

Figure 5.24. An interesting pattern mined from a real data set, where T stands for transportation and HM stands for Home Management Services.

who are in the foster care system after getting a substantiated report of abuse and neglect have very similar service patterns. Within one month of the report, there is an investigation and the child is put into foster care. Once children are in the foster care system, they stay there for a long time. This is consistent with the policy that all reports of abuse and neglect must be investigated within 30 days. It is also consistent with our analysis on the length of stay in foster care.

Interestingly, when a conventional sequential algorithm is applied to this data set, variations of this consensus pattern overwhelm the results, because roughly half of the sequences in this data set followed the typical behavior approximately.

The rest of the sequences in this data set split into clusters of various sizes. One cluster formed around the few children (57 sequences) who have short spells in foster care. The consensus pattern was

$$\langle (RPT)(INV, FC)(FC)(FC) \rangle.$$

There were several consensus patterns from very small clusters with about 1% of the sequences. One such pattern of interest is shown in Figure 5.24.

There are 39 sequences in the cluster. Our clients were interested in this pattern because foster care services and home management services were expected to be given as an "either/or" service, but not together to one child at the same time. Thus, this led us to go back to the original data to see if indeed many of the children received both services in the same month. Our investigation found that this was true, and lead our client to investigate this further in real practice. Was this a systematic data entry error or was there some components to Home Management Services (originally designed for those staying at home with their guardian) that were used in conjunction with Foster Care Services on a regular basis? Which counties were giving these services in this manner? Such an important investigation would not have been triggered without our analysis because no one ever suspected there was such a pattern. It is difficult to achieve the same results using the conventional sequential analysis methods because when the support threshold is set to $min\_support = 20\%$, there is more than $100,000$ sequential patterns and the users just cannot identify the needle from the straws.

Notes

1 We will define shortly that a pattern P is a super-pattern of P' if P' can be obtained by dropping some symbol(s) in P. In such a case, P' is also called a subpattern of P.

2 Note that P' does not have to be a contiguous portion of P. Gaps are allowed and an upperbound or lowerbound on the gap length can be imposed if desired.

3 In bio-informatics, a score matrix (such as BLOSUM 50) [7] is derived to indicate the likelihood of two amino acids coming from the same origin for evaluating pattern similarity. The compatibility matrix can similarly be derived.

4 This is because a subsequence does not have to be a contiguous portion in a sequence.

5 The spread of a random variable is defined as the difference between the maximum possible value and the minimum possible value of the random variable.

6 By definition, the match of a pattern in the sample data is the average match of every sample.

7 Each sequence consists of dozens to thousands of amino acids with an average length of around 500.

References

[1] Agrawal, R., and Srikant, R. (1995). Mininig sequential patterns. *Proc. of the Int'l Conference on Data Engineering (ICDE)*. pp. 3-14.

[2] Ayres, J. Flannick, J., Gehrke, J., and Yiu, T. (2002). Sequential pattern mining using a bitmap representation. *Proc. of the Eighth ACM Int'l Conference on Knowledge Discover and Data Mining (KDD)*. pp. 429-435.

[3] Baeza-Yates, R., and Navarro, G. (1999). Faster approximate string matching. *Algorithmica*. 23(2):127-158.

[4] Bayardo, R. (1998). Efficiently mining long patterns from databases. *Proc. ACM SIGMOD Int'l. Conference on Management of Data (SIGMOD)*. pp. 85-93.

[5] Coggins, J. (1983). Dissimilarity measures for clustering strings. *Time Warps, String Edits, and Macro-molecules: the Theory and Practice of Sequence Comparison*. pp. 253-310.

[6] Domingos, P., and Hulten, G. (2000). Mining high-speed data streams. *Proc. of the Sixth ACM Int'l Conference on Knowledge Discover and Data Mining (KDD)*. pp. 71-80.

[7] Durbin, R., Eddy, S., Krough, A., and Mitchison, G. (1998). *Biological Sequence Analysis: Probabilistic Models of Proteins and Nucleic Acids*. Cambridge University.

[8] Fukunaga, K. and Narendra, P. (1975). A branch and bound algorithm for computing k-nearest neighbors. *IEEE Transactions on Computers*. 24:750-753.

[9] SAS. (2000). Proc Modeclust. SAS/STAT user guide. *SAS Online Document.*

[10] Garofalakis, M., Rastogi, R., and Shim, K. (1999). SPIRIT: sequential pattern mining with regular expression constraints. *Proc. of Int'l Conference on Very Large Databases (VLDB),* pp. 223-234.

[11] Gusfield, D. (1997). *Algorithms on Strings, Trees, and Sequences: Computer Science and Computational Biology.* Cambridge University Press.

[12] Han, J., Pei, J. and Yin, Y. (2000). Mining frequent patterns without candidate generation. *Proc. SIGMOD Int'l. Conference on Management of Data (SIGMOD),* pp. 1-12.

[13] Han, J., Pei, J., Mortazavi-Asl, B., Chen, Q., Dayal, U., and Hsu, M. (2000). FreeSpan: frequent pattern-projected sequential pattern mining. *Proc. of the Sixth ACM Int'l Conference on Knowledge Discover and Data Mining (KDD).* pp. 355-359.

[14] Hoeffding, W. (1963). Probability inequalities for sums of bounded random variables. *Journal of the American Statistical Association.* (58):13-30.

[15] Koontz, W., Narendra, P. and Fukunaga, K. (1975). A branch and bound clustering algorithm. *IEEE Transactions on Computers.* 24(9):908-915.

[16] Kum, H., Pei, J., Wang, W., and Duncan, D. (2003). ApproxMAP: approximate mining of consensus sequential patterns. *Proc. of SIAM Int'l. Conference on Data Mining (SDM).*

[17] H. Mannila and H. Toivonen. (1997). Levelwise search and borders of theories in knowledge discovery. *Data Mining and Knowledge Discovery.* 1(3):241-258.

[18] Mannila, H., Toivonen, H., and Verkamo, A. (1997). Discovery of frequent episodes in event sequences. *Data Mining and Knowledge Discovery Journal,* 1(3):259-289.

[19] McPherson, G., and DeStefano, S. (2002). *Applied Ecology and Natural Resource Management.* Cambridge University Press.

[20] National Center for Biotechnology Information. Available at "http://www.ncbi.nlm.nih.gov".

[21] Pasquier, N., Bastide, Y., Taouil, R., and Lakhal, L. (1999). Discovering frequent closed itemsets for association rules. *Proc. Seventh Int'l. Conference on Database Theory (ICDT).* pp. 398-416.

[22] Pei, J., Han, J., Mortazavi-Asl, B., Pinto, H., Chen, Q., and Dayal, U., and Hsu, M. (2001). PrefixSpan: mining sequential patterns by prefix-projected growth. *IEEE Int'l Conference on Data Engineering (ICDE).* pp. 215-224.

[23] Srikant, R., and Agrawal, R. (1995) Mining generalized association rules. *Proc. of the Int'l Conference on Very Larg Databases (VLDB).* pp. 407-419, 1995.

[24] Srikant, R., and Agrawal, R. (1996). Mining sequential patterns: generalizations and performance improvements. *Proc. of Int'l Conference on Extended Database Technologies (EDBT).* pp. 3-17.

[25] Toivonen, H. (1996). Sampling large databases for association rules. *Proc. of the Int'l Conference on Very Larg Databases (VLDB).* pp. 134-145.

[26] Vitter, J. (1987). An efficient algorithm for sequential random sampling. *ACM Transactions on Mathematical Software*. 13(1):58-67.

[27] Yang, J., Wang, W., and Yu, P. (2001) Mining long sequential patterns in a noisy environment. *IBM Research Report*.

[28] Yang, J., Wang, W., Yu, P., and Han, J. (2002). Mining long sequential patterns in a noisy environment. *Proc. ACM SIGMOD Int'l. Conference on Management of Data (SIGMOD)*. pp. 406-417.

[29] Zaki, M., Parthasarathy, S., Ogihara, M., and Li, W. (1997). Parallel algorithm for discovery of association rules. *Data Mining and Knowledge Discovery*, 1:343-374.

Chapter 6

CONCLUSION REMARK

In summary, we share our experiences in sequential pattern mining and present several representative models developed to accommodate skewed symbol distributions and potential noises in the form of insertion and substitution. We also experimentally demonstrate that these models are superior than previous ones in several real applications. However, they are by no means perfect and complete. In fact, this is still an active research area with many unsolved challenges. For instance, the order of symbols in the sequence may be disrupted by noises. This phenomenon is not considered by any method we present in this book. Another example of such challenges is how to cope with the compounded effect of multiple noise types. There will be a long journey before we can claim the study of sequential patterns concluded.

Index